The Integration of Alternative
Information Systems: An Application
to the *Hogs and Pigs* Report

CARD Monograph 92-M4

The Integration of Alternative Information Systems: An Application to the *Hogs and Pigs* Report

Karl Durwood Skold

Center for Agricultural and Rural Development
Iowa State University
Ames, Iowa 50011
U.S.A.

Karl D. Skold is a commodity analyst with Quaker Oats Company, Chicago, Illinois. He received the Ph.D. in agricultural economics from Iowa State University and is a former CARD research associate.

This monograph is based on a dissertation study conducted while the author was at CARD. Members of the doctoral committee were Stanley R. Johnson, Charles F. Curtiss Distinguished Professor of Agriculture and director of CARD; Gene A. Futrell, professor of economics; J. Arne Hallam and Peter F. Orazem, associate professors of economics; and Wayne A. Fuller, Distinguished Professor of Sciences and Humanities and professor of statistics.

Copyright 1992. Center for Agricultural and Rural Development
Iowa State University, Ames, Iowa 50011

Library of Congress Catalog No. 92-072452
ISBN 0-936911-03-4

CONTENTS

FIGURES

TABLES

FOREWORD

This CARD monograph presents results that have important implications for the development of data on the livestock subsector for U.S. agriculture. Specifically, the analysis shows the importance of integrating analytical systems and biological restrictions with available information that is now summarized in the series reported publicly by the National Agricultural Statistics Service (NASS). In general, the conclusions of the monograph support the idea that alternative sources of information, including that on the biology of swine reproduction, should be included and could improve the quality of U.S. aggregate-reported statistics.

The analytical model developed in this monograph begins first with biological relationships, showing the importance of these restrictions in determining the supply response for swine. Modeling of economic incentives for producers is then added. Interestingly, the economic signals provided by standard supply response mechanisms add information, but are dominated by biological responses. This is quite in contrast to models that have been developed that rely entirely on economic incentives and include lagged prices of inputs and outputs as major determinants of supply response.

The analysis is extended from supply to include demand response. Results show that the information that can be added to survey-generated data should reflect not only the biology of swine production and the structure of supply, but demand as well. Demand for livestock products in the United States is according to available empirical results and, as indicated by the empirical results presented in this monograph, is itself dynamic in nature. Thus, the supply response for swine reflects a cascading of biological, economic incentives on the supply side, and economic incentives on the demand side. And, the latter two are highly dynamic in nature.

This study demonstrates the value of augmenting existing information systems with results from models that reflect not only the past correspondence of the information and forecasting systems to realizations but reflect, as well, information from biology and economic structure. This augmentation adds importantly to the data generated from surveys.

Stanley R. Johnson
Director,
Center for Agricultural
and Rural Development

PREFACE

Federal budget cuts in the late 1980s reduced the sample survey coverage of U.S. Department of Agriculture (USDA) crop and livestock reports. These cuts renewed concerns about the adequacy and reliability of the reports, in general, and of the *Hogs and Pigs* report in particular. The object of this study is to demonstrate that more consistent and reliable initial estimates of key hog-supply and inventory categories can be generated by augmenting the survey-based *Hogs and Pigs* report estimates with market information. Two alternative pork-sector econometric models with alternative expectation regimes are proposed as a relatively costless means to expand the information set on which the *Hogs and Pigs* report estimates are based.

The first econometric model is dynamic and nonlinear and incorporates the rational expectation hypothesis with forward-looking expectations. The second incorporates futures-market price expectations. Breeding herd decisions of pork producers are based on distant closing futures prices of live hogs and corn. Both econometric models integrate restrictions based on the biological processes of pork production as prior information in their supply components.

The market information from the two econometric models is synthesized in one-step-ahead forecasts of key hog-supply and inventory categories. These model-based forecasts are combined with the USDA initial estimates by using alternative composite forecasting techniques. The results suggest that including market information in the data evaluation and estimation procedures can reduce errors in the USDA initial estimates. Market information often compensates for errors in the USDA estimates. Thus, econometric models and composite-forecasting techniques may provide a viable, cost-effective means to improve the consistency and reliability of the initial Hogs and Pigs report estimates.

The decision to come back to the place of my birth—Ames—was a good one. My four-plus years at Iowa State have been very rewarding. My major professor, Dr. Stanley R. Johnson, deserves much of the credit. His enthusiasm for life and economics is contagious, and I thank him for the many opportunities he has opened up for me. Stan has fostered a truly unique, productive environment for graduate study at the Center for Agricultural and Rural Development (CARD).

I also thank the rest of my committee members: Dr. Gene A. Futrell for his considerate, helpful nature and for his insight into the livestock markets; Drs. J. Arne Hallam and Peter F. Orazem for their interest and advice; and Dr. Wayne A. Fuller for his inquisitiveness and intuition.

Finally, I thank my wife, Molly. During the dissertation process, I learned as much about her as she learned about hogs and pigs. She continues to amaze me with her continual love, patience, encouragement, and support. I look forward to our life together because I know it always will be filled with wonder and surprise.

Karl D. Skold

Chapter 1

Introduction

The demand for information is intrinsically tied to uncertainty. Gathering information can be thought of as a process by which individuals lessen uncertainty in their decision making (Hirshleifer 1973). The nature of agricultural markets fosters a demand for information. Agricultural markets are inherently variable. Production decisions are necessarily made months before crops or livestock reach their markets. Weather, pests, the actions of domestic and foreign governments, and assorted random factors contribute to uncertainty during production periods. Also, the inelastic demand for most agricultural products induces further market instability. Modest supply movements can create substantial changes in farm-level prices. Finally, the atomistic structure of agricultural production generates instability. Individual producers may act as if their production decisions do not affect market prices, and this in turn contributes to swings in the level of production.

With these inherent uncertainties in agricultural markets, the demand for information comes from a wide range of sources--from government agencies to individual producers. Government has entered agricultural markets to stabilize prices and producer returns. To achieve these and other objectives, the government's need for accurate policy assessment has expanded the demand for data and information. At the individual producer level, the demand for information has evolved as the ability to assimilate information into decision making processes has advanced by way of rapid improvements in communication and computer technology. The proliferation of personal

1

computers and software programs provides ease in processing raw data into usable information for decisions (King 1983). Agricultural producers can now instantaneously access price quotes from major agricultural markets, obtain analysis and advice on price trends, and even follow worldwide developments in agricultural markets and in the weather (Just 1983).

The increase in the demand for information has expanded the volume and dissemination of information. The development of futures markets for agricultural commodities has provided market information and a price discovery mechanism for producers (Hieronymus 1971). Other private sources of market information also have developed, such as Doanes Agricultural Service, Sparks Commodities, Inc., and the WEFA Group. The information from these commercial firms is supplemented by newsletters, trade magazines, and information services such as Future Source. Also, state and federal extension personnel release market information through the media, bulletins, presentations, and computer networks.

The expansion in the demand for data and information is constrained by the underlying agricultural database. The characteristics of information limit the incentive for individual data collection. Collecting market data is time and labor intensive, a costly activity (Simpson 1966; Bonnen and Nelson 1981). Economies of scale are found in the production of data and information (Tullock 1970). The public good characteristics of information, nonrivalry, and nonexcludability provide further disincentives for individual collection. Often, privately held information is freely transmitted once it is incorporated in the decisions of market participants. This is particularly true in agricultural markets with widely disseminated cash and futures market prices. Also, the timeliness of information makes the creation of information a risky endeavor. These factors cause underinvestment in the production of information by the private sector (Arrow 1962). Consequently, the main supplier of primary data about agricultural markets is the public sector.

U.S. Agricultural Data Base Issues

The principal supplier of data and information on the current disposition of the domestic agricultural sector is the U.S. Department of Agriculture (USDA) (see Eisgruder 1973; Upchurch 1977; and Trelogan et al. 1977). The collection and dissemination of agricultural statistics by the U.S. government, however, preceded the creation of the USDA. In 1839, the U.S. Patent Office began collecting agricultural statistics to "aid farmers in marketing their crops and enable them to gain some of the profits going to speculative monopolists" (Koffsy 1962, 1755). The USDA, which began supplying crop and livestock statistics in 1862 (Helmberger et al. 1981, 566), continues to orient the agency information system toward the needs of producers (Upchurch 1977).

Producers and other agribusinesses use crop and livestock estimates produced by the USDA in making production, storage, and marketing decisions. The USDA, through its National Agricultural Statistics Service (NASS, formerly called Statistical Reporting Service), is the primary source of commodity supply estimates. Annually, NASS issues nearly 300 reports. These contain forecasts and estimates of crops, livestock, poultry, dairy, prices, farm labor, weather, and related agricultural inputs.

Primarily, the estimates are a result of extensive sampling of agricultural producers. Producers voluntarily provide data about their crop and livestock enterprises. The scope and quality of this information base increased with continued refinements in sampling techniques and data analysis, combined with advances in computer technology (Upchurch 1977; Trelogan et al. 1977). The reports have become progressively more accurate, timely, and accessible. They are considered the benchmark of supply estimates for most agricultural commodities in the United States.

Nevertheless, with the changing nature of the agricultural sector, the adequacy and reliability of agricultural data have continually been reassessed. Needs for broadening the scope of the current data system have developed as the structure of agriculture and the policy agenda have shifted (Bonnen 1977). Critics of the current data system have focused on its inability to monitor the changing structure of agriculture and adequately depict the welfare of rural economies (Bottum and Ackerman 1958; Cochrane 1966; AAEA 1972; Bonnen 1977; Bonnen and Nelson 1981).

Recent cuts in the funding that maintains the agricultural data base have renewed concerns about its adequacy to provide timely and accurate information for individual and policy decision making (Bonnen 1983; Gardner 1983; Just 1983). In March 1982, budgetary reductions forced the elimination of 26 USDA commodity reports (USDA 1982). Some data series were suspended, and the frequency of selected reports and estimates was reduced. The survey coverage of some reports also was reduced.

The budget reductions, in part, result from pressures to expand the coverage of the existing database, both in agricultural and nonagricultural sectors of the economy. These pressures expand the usefulness of the agricultural database beyond providing relevant market information to producers and undoubtedly will persist. Increased allocations to crop and livestock statistics seem doubtful, given the current fiscal constraints. Also, if the attitude prevails that more effort is devoted to "improving data about hogs than about rural people and their welfare" (Upchurch 1977, 309), scarce dollars will undoubtedly be reallocated elsewhere.

Livestock Report Issues

The livestock reports that have received the most critical attention are the *Cattle on Feed* and *Hogs and Pigs* reports. These two reports are the primary source of short-term supply information for fed cattle and hogs, respectively. Initially, criticism of these reports focused on their informational deficiencies. Improving their timeliness, content, coverage, and accuracy were deemed necessary (Kutish 1955; Ives 1957; Luby 1957; Ferris 1962). Suggestions included distinguishing cattle and hog slaughter by sex and providing monthly pig crop estimates.

The USDA responded positively, in general, to these demands for expanded coverage. For example, quarterly survey coverage of the *Hogs and Pigs* report was expanded from six states in 1954 to fourteen states in 1973 (USDA 1961; 1973). With the 1982 budget reductions, however, the survey coverage for the livestock reports was reduced. Coverage of the *Cattle on Feed* report was reduced from twenty-three to thirteen states, and the *Hogs and Pigs* coverage was reduced from fourteen to ten states (USDA 1982).

The reduction in sample coverage combined with apparent inconsistencies in the livestock reports has renewed concerns about the accuracy and reliability of the livestock reports (USDA 1988). Accurate supply estimates are particularly important in the livestock markets. Among other factors, the biologically based sequential nature of cattle and hog production and the long production period foster this need. Producers must assess the profitability of increasing the size of their breeding herds. On the aggregate level, this requires forfeiting current supply to increase future production capacity. Because the livestock reports indicate the current productive capacity (breeding herd size) and the supply flowing to the market (cattle on feed, market hogs), the reports have a direct influence on producers' price expectations and, in turn, their investment decisions.

Livestock production does not receive direct price support from government programs, so there is not substantial government intervention in supply determination. The livestock reports are a fundamental feature of price determination because they provide the most reliable estimate of supply.

Hence, accurate reports contribute to more informed price and profitability assessments in livestock producers' planning.

Hogs and Pigs Report Issues

The *Hogs and Pigs* report has continued to receive the bulk of criticism (Kutish 1955; Luby 1957; Ferris 1962; *Futures* 1984; Hohmann 1987; USDA 1988). The report, released quarterly, provides the primary indication of the near-term hog supply. Included in the report are estimates of the breeding herd, sows farrowing, market hogs, and the size of the pig crop. Market hogs are further subdivided into weight categories. Estimates of producer farrowing intentions also are included in the report.

Many producers are concerned about possible impacts of the report on market prices. Specifically, "there's a nagging suspicion among hog producers that the quarterly *Hogs and Pigs* reports only make prices go down" (USDA 1977b, 2). Miller (1979) and Hoffman (1980) have demonstrated that this perception is unwarranted and that hog markets react to the supply of new information according to basic economic principles. Discrepancies between beliefs before and after the report do not constitute unreliable estimates; instead, they may suggest insufficient information on market conditions.

Another concern is that apparent inconsistencies exist in the report (*Futures* 1984; Hohmann 1987; USDA 1988). Blanton et al. (1985) contend that periodic and systematic errors may exist in the breeding herd inventory numbers. This contention was supported with a cursory analysis of breeding herd estimate errors. The breeding herd is the fundamental determinant of the pork supply. Given the regularities in sows farrowing, the size of the breeding herd should closely correspond with the level of slaughter approximately six months later. Five to six months are required to wean and then feed a pig to slaughter weight. Using this simple biological relationship, Blanton et al. (1985) compared percentage changes in the breeding herd data with the percentage changes in the subsequent slaughter data.

In general, the percentage change in the breeding herd did correspond to movements in the slaughter. Some outliers existed, but could be explained by changes in farrowings, weather, feed conditions, and a host of other factors. Nevertheless, Blanton et al. proposed a biologically based econometric model of the pork supply, combined with the sample-based estimates, to reduce apparent outliers in the breeding herd data. This composite forecast would have an error variance no greater than the smallest of the individual forecasts (Bates and Granger 1969; Granger and Newbold 1986) and, thus, would always outperform either individual forecast. Blanton et al. (1985) maintain that the actual level of the breeding herd lies between the model estimate and the USDA estimate.

Meyer and Lawrence (1988) examined the accuracy of the *Hogs and Pigs* report with a tracking model of hog production. Their model also incorporated information on the biology of hog production. They included information such as gilt retention, average death loss, and average daily gain. Then, they traced slaughter data back to the implied pig crop and categories of market hogs by using the production information. The implied estimates for the supply categories were compared with initial estimates in the *Hogs and Pigs* report.

Their results suggest that a high hog-corn price ratio preceding the release of USDA estimates results in an upward bias in the USDA estimated pig crop. This implies that current market conditions may influence producers' survey responses. Their results also suggest an upward bias in the USDA estimates of the second-quarter pig crop and market hogs between 120 and 179 pounds, and an overemphasis of seasonality in the supply estimates. However, Meyer and Lawrence (1988) did not validate their model by comparing the hog-supply estimates from their tracking model with the final estimates of the supply and inventory categories. Thus, their results may depend on the biological assumptions made and on other variables, such as weather and feed conditions, that are not incorporated in the tracking model.

Errors in the *Hogs and Pigs* report are discovered only when subsequent slaughter does not correspond with previous estimates of supply and inventory. These errors may be attributed to many factors. Sampling and nonsampling errors are prevalent in all survey-based estimates. Nonrespondents are always a source of error, a result of voluntary participation by producers and associated selectivity bias. The current sampling techniques and sampling frames may not given an adequate representation of pork producers' behavior.

Concerns about the adequacy of estimates in the *Hogs and Pigs* report may reflect the tremendous structural change in the pork industry. Pork production, packing, and marketing have become increasingly concentrated (Hayenga et al. 1985). With the increase in concentration and capital intensity in the pork industry, market participants have adopted advanced management and information systems. Updating samples, a costly procedure, may not be completed with sufficient frequency to capture the behavioral changes induced by a more concentrated and integrated pork sector. The current USDA hog supply and inventory estimates may not be released often enough or be sufficiently accurate to meet the informational demands of the sector.

Proposed Solutions

Because of the apparent inconsistencies in the *Hogs and Pigs* report, and cost constraints, it is unlikely that estimates will be improved by expanding sample coverage or by updating samples and releasing reports more frequently. Expanding the informational content by including the sex of slaughter also is unlikely because of the proprietary nature of these data (Blanton et al. 1985). Bonnen and Nelson (1981, 343) suggest that improvements may be found in adopting "statistical strategies and methods that can substitute for the more expensive conventional survey and census methods."

As illustrated by Blanton et al. (1985), econometric models can expand the information set on which the estimates are based. Incorporating information from econometric models may render more accurate estimates in the *Hogs and Pigs* report and reduce inconsistencies among the reports. Econometric models provide a representation of industry behavior and allow for the incorporation of biological and physical restrictions that govern the hog production process. The biological restrictions afford added integrity in the relationship between short-term supply movements and the long-term structure of supply response.

The supply response of hogs is not entirely governed by the biology and physical structure of the hog production process. Producers have discretion in their production planning decisions. Producers adjust gilt retention, marketing plans, and feeding practices, for example, in response to changing economic conditions and perceptions. The manner in which individuals process information is unobservable. Nevertheless, in econometric modeling, the accurate depiction of an individual's ability to process information forms the basic reliable representation of industry behavior (Chavas and Johnson 1982).

Typically, individuals' expectations are represented as a distributed lag of relevant decision variables. These forms have been criticized because they imply a waste of information (Lucas 1976). Another approach, the rational expectation hypothesis (REH), as developed by Muth (1961), states that individuals value information, which is typically scarce, and use it efficiently in determining their future economic activities. The REH forces a consistency between the structural representation of the industry and the expectation mechanism used by individuals participating in the system.

Futures market price quotations also can be used to represent the unobservable expectations of market participants (Chavas and Johnson 1982). Expectations based on futures market prices can be justified as an efficient use of information, implied by the REH. If expectations are rational, no reason exists for futures market participants to have different expectations than nonparticipants (Gardner 1976). Futures market prices have been implemented in empirical estimates of agricultural crop-supply response by Gardner (1976), Chavas, Pope, and Kao (1983), and Subotnik and Houck (1982), and in livestock supply estimates by Miller and Kenyon (1980).

Econometric models can expand the information set used to develop the *Hogs and Pigs* report estimates, and rational expectations and expectations based on futures market prices can be used to generate the unobserved anticipations of market participants. Thus, the pork industry models combine information about the biological processes that govern hog production and alternative mechanisms that describe how producers process market information. This incorporates additional information on behavioral aspects of the industry and known growth constraints in pork production.

By combining USDA survey sample estimates and the econometric model estimates in the data evaluation process, the USDA might produce better indications of the movements in the hog supply. Composite forecasting techniques can be used to combine the independent USDA and model estimates into a single prediction that outperforms its individual components (Bates and Granger 1969; Johnson and Rausser 1982; Granger and Newbold 1986). Bessler and Brandt (1979), Brandt and Bessler (1981; 1983), and others have demonstrated this improved forecast precision by applying composite forecasting techniques in agricultural markets.

There is a need for improved *Hogs and Pigs* report estimates that can be obtained without resorting to more costly survey procedures. Structural econometric models that use different information bases than USDA survey estimates may provide a viable solution.

Although the conceptual basis for improving the accuracy of the *Hogs and Pigs* report is clear, the value of improving the estimates is not. For example, Hayami and Peterson (1972) found that there are considerable social gains in reducing the USDA forecast errors, and that a significant underinvestment exists in providing commodity supply information. However, later results suggest that this is not certain. The value of improving USDA estimates may depend on the supply and demand conditions that exist (Bullock 1976; 1981), the ability of individuals to process new information (Bradford and Kelejian 1977; 1978), and whether market participants forecast more accurately than the government (Falk and Orazem 1986). Thus, the value of commodity forecasts in general, and the hog market in particular, are not readily apparent.

Objectives

The general objectives of this study are to develop and incorporate alternative information systems to assist in the data evaluation and estimation procedures of the *Hogs and Pigs* report. Specifically, this study incorporates market information from both rational expectations and futures market expectations models of the pork sector into the determination of the initial estimates in the *Hogs and Pigs* report. The econometric models of the pork industry with alternative expectation mechanisms may provide an alternative information source to replace recent losses in survey coverage. Combining the survey estimates with the model's predictions may provide more accurate and consistent estimates. Estimates from the econometric models are relatively inexpensive sources of additional information compared to expanded survey coverage. This latter feature is of interest, given the current state of fiscal austerity.

The specific objectives are:

- Develop an econometric model of the pork sector in which market participants' expectations are rational and the supply structure is consistent with the hog growth process.

- Develop an econometric model that uses futures market price expectations and is consistent with the biological constraints of hog production.

- With composite forecasting techniques, combine predictions of the econometric models with survey estimates to provide more accurate and consistent estimates of key hog supply indicators.

Chapter 2

The Development
of the *Hogs and Pigs* Report Estimates

The U.S. government traditionally has been the major source of domestic crop and livestock supply estimates. These supply estimates rely on data from surveys of crop and livestock producers. In part, the changes in survey and sampling methods employed in deriving the hog supply estimates reflect changes in the structure of the livestock sector in general and the pork sector in particular. Advances in statistical survey methods and data processing abilities have improved the timeliness and precision of the estimates.

The pork sector has undergone a dramatic structural transformation. Since the inception of USDA estimates of the hog supply, the pork industry has progressively shifted from an industry with many small homogenous producers to an industry with production concentrated among larger producers. These large producers have decreased costs and increased efficiency by adopting capital-intensive confinement units and advanced management practices (Van Arsdall and Nelson 1984). Fewer farms maintain hog production enterprises, as shown in Table 2.1. Hog production, once an ubiquitous enterprise, has become more concentrated and specialized.

The trend of an increasing percentage of total production originating from producers with large capacity is depicted in Table 2.2. In 1964, farms with greater than 10,000 head sales accounted for less that 8 percent of the total number of hogs and pigs sold (U.S. Bureau of the Census 1967), and less than

Table 2.1. Number of farms and percentage of total farms reporting hog and pig inventories

Year	Number of Farms	Percent of Total Farms
1900	4,335,363	75.6
1910	4,351,751	68.4
1920	4,850,807	75.2
1930	3,535,119	56.2
1940	3,766,675	61.8
1950	3,011,807	55.9
1959	1,846,982	49.8
1969	686,097	25.1
1978	512,292	20.7

SOURCE: U.S. Bureau of the Census 1967, 1978, 1981.

Table 2.2. Number of farms selling hogs and pigs by size group, 1959-82

Year	Total farms	Number sold annually (thousands) per farm				
		1-99	100-199	200-499	500-999	>10,000
1959	273.4	1018.7	161.6	81.6	10.0	1.5
1964	802.6	547.2	139.2	94.7	17.4	4.1
1969	604.2	361.3	109.4	101.5	25.4	6.6
1974	449.8	260.3	75.5	77.0	26.1	10.8
1978	423.5	237.4	67.5	73.1	29.7	15.8
1982	315.0	163.1	44.4	55.9	30.0	21.6

SOURCE: U.S. Bureau of the Census 1967, 1978, 1981, 1984.

twenty years later, in 1982, accounted for nearly 50 percent of the total number of hogs and pigs sold (U.S. Bureau of the Census 1984).

Changes in the structure of the pork industry are reflected in changes in the USDA survey sampling techniques. The survey sampling procedures for livestock in general, and hog production in particular, have shifted to methods more suitable for sampling populations that have more variable characteristics such as size of production enterprise. This chapter details the changes in the survey coverage and sampling methods the USDA has used to obtain timely hog supply and inventory estimates.

Review of the Hog Supply and Inventory Estimates

The *Hogs and Pigs* report provides quarterly estimates of the composition and movements in U.S. hog supply and inventory on March 1, June 1, September 1, and December 1. The report includes estimates of inventory level, both market and breeding hogs, number of sows farrowed, farrowing intentions, and the pig crop. The market hogs inventory estimate is subdivided into four weight categories: less than 60 pounds, 60 to 119 pounds, 120 to 179 pounds, and more than 180 pounds. The December report also includes estimates of the number and size disposition of hog operations in the United States.

The U.S. government has revised the content and scope of these hog supply and inventory estimates, as well as the survey and sampling procedures. The objective has been to provide a more complete and accurate depiction of movements in hog supply and inventory. The report has been changed because of budget constraints and reductions. In Table 2.3 the chronology of U.S. government efforts to provide commodity supply estimates in general, and hog supply and inventory estimates in particular, is shown.

The collection of agricultural statistics preceded the creation of the USDA in 1862. In 1839, Congress appropriated $1,000 to the U.S. Patent Office for the collection of agricultural statistics and distribution of seed (Koffsy 1962; USDA 1983). In 1866, after the creation of the USDA, releases of annual supply estimates for major crops and livestock were established (USDA 1983), based on interpolating annual data for years between enumeration as part of the Census of Agriculture.

In 1922, the USDA published the percentage distribution of the births of pigs (USDA 1947). This monthly pig crop survey was conducted by rural mail carriers who distributed questionnaires to producers on their routes. In 1924, the monthly distribution of farrowings was added to the survey. These

Table 2.3. Chronology of the collection and reporting of agricultural statistics

Year	Development
1839	Congress appropriates $1,000 to the Patent Office for the collection of agricultural statistics and the distribution of seed.
1862	U.S. Department of Agriculture established.
1863	Division of statistics formed in the USDA.
1866	Beginning of continuous series of annual reports on supply estimates of major crops and livestock.
1922	Rural mail carriers distribute survey questionnaires on the monthly distribution of the pig crop.
1924	Estimates of the monthly distribution of sows farrowing added to the pig crop survey.
1929	Estimates of the level of fall and spring farrowings established for the United States and the Corn Belt.
1954	Quarterly estimates of hog supply and inventory numbers for Indiana, Illinois, Wisconsin, Minnesota, Iowa, and Kansas established.
1956	Ohio, Missouri, and South Dakota added to the quarterly survey of hog supplies and inventories.
1957	Quarterly hog survey expanded to ten states by the addition of Nebraska.
1967	Probability sampling principles initiated for the livestock surveys.
1970	Multiple-frame survey methods adopted for the cattle- and hog-supply and inventory estimation procedures.
1973	Hog survey expanded to fourteen states. Georgia, Kentucky, Texas, and North Carolina added to the quarterly survey.
1982	Survey coverage reduced to ten states. Georgia, North Carolina, Indiana, Illinois, Missouri, Iowa, Minnesota, Kansas, and Nebraska included in the quarterly survey.

SOURCE: USDA 1983, 1988.

surveys provided the first timely estimates on the supply of hogs, as well as an industry performance measure, pigs saved per litter.

In December 1929, rural mail carriers conducted the survey that produced estimates of the number of sows farrowing in the fall and in the spring. Aggregate estimates were made for the United States and detailed estimates were provided for the Corn Belt—Ohio, Indiana, Illinois, Michigan, Wisconsin, Minnesota, Iowa, Missouri, North Dakota, South Dakota, Nebraska, and Kansas—were established in 1954 (USDA 1961). In 1956 Ohio, Missouri, and South Dakota were added to the quarterly survey. In 1957, a tenth state, Nebraska, was added to the quarterly survey of hog supplies and inventories. The ten states in the quarterly hog surveys are depicted in Figure 2.1.

Figure 2.1. Hog survey coverage, 1957 to 1972

The survey was expanded to fourteen states in 1973 (USDA 1977a). In addition to the original ten states, Georgia, Kentucky, Texas, and North Carolina were added to the quarterly survey. The fourteen states in the survey beginning in 1973 are depicted in Figure 2.2. The June and December reports provided aggregate estimates for the entire United States as well as detailed estimates for the fourteen major producing states. The March and September reports included estimates of only the fourteen major producing states.

Figure 2.2. Hog survey coverage, 1973 to 1982

The hog survey coverage later was reduced, beginning with the 1982 June report. The survey coverage was reduced to ten states—Georgia, Illinois, Indiana, Iowa, Kansas, Minnesota, Missouri, Nebraska, North Carolina, and Ohio. These states included in the hog survey from 1982 on are highlighted in Figure 2.3. The estimate in the June report was based on a survey of the ten states, plus a composite estimate for the forty remaining states (USDA 1984).

Figure 2.3. Hog survey coverage, 1982 to present

Sampling and Survey Procedures

The survey procedures also have been refined as the USDA attempted to provide a more accurate depiction of the livestock economy. Again, initial annual estimates of hog supply and inventory numbers were based on interpolations of annual data from years between enumeration as part of the Census of Agriculture. Beginning in the 1920s and continuing through the 1960s, the estimates of movements in the hog supply were developed from the survey questionnaires dispersed by rural mail carriers. The survey method was not based on systematic statistical sampling principles. Surveys were enumerated from producers that "were believed to be well informed, and would report regularly" (USDA 1983, 23). Thus, selectivity bias was present in the survey. Nevertheless, these mail surveys were cost effective and provided a fairly accurate means of gathering data because the population was relatively homogenous in production characteristics and regional location.

Beginning in 1967, survey procedures based upon probability sampling principles were initiated for livestock surveys (USDA 1983). The diversity of the population of livestock producers had evolved with changes in industry structure. Variations in production practice, operation size, and geographical location became important characteristics in differentiating producers. Mail surveys using lists of well-informed and dependable respondents became less reliable because respondent and nonrespondent farms were likely to be dissimilar.

Current Survey and Sampling Procedures

In 1970, multiple frame sampling techniques were instituted for the livestock surveys. The multiple frame sampling technique consisted of obtaining samples from both area and list frames. A frame is the basic list or reference that defines every element in the population from which the sample is taken (Stopher and Meyburg 1979). Thus, the area frames are used for random sampling from selected areas or tracts of land. The list frame method samples from lists of producers within a region. Combining the area and list frame sampling methods provides a more reliable estimate of total hog inventory and, thus, a better indication of producer behavior.

Area Frame

The area frame survey technique derives samples from randomly selected land tracts. The land tracts, termed *segments*, include all land in the designated area. These segments must cover the land area completely, and have no more than one chance of being selected. Consequently, the area frame sample is complete. That is, all land segments have a known chance of being selected, and hence all items of interest have a chance of being selected within the sampled segment. These segments vary inversely in size to intensity of land use, and average about one square mile in major agricultural producing areas (USDA 1983). The total land actually surveyed in the area frame sample is about .5 percent of the U.S. total land area (USDA 1983).

Boundaries of the land area are defined and partitioned into identifiable land segments, which are then stratified according to current land use. Stratification of the area frame increases the precision of the estimates and allows for increased sampling of specific land-use areas. Thus, stratification ensures that agricultural regions are adequately represented in the survey. Table 2.4 gives an example of stratification categories within the area frame. Stratification categories include percentage of land under cultivation, rangeland, and urban (USDA 1988). Stratification is conducted at the state level.

Table 2.4. Example of area frame stratification by land use

Stratum	Population	Sample	Sampling Rate
75% cultivation	25,062	170	147
50-74% cultivation	21,736	120	181
15-49% cultivation	21,284	100	213
Agri-urban	3,091	14	221
Urban	2,941	12	245
Rangeland	3,163	15	211
Nonagricultural	321	4	80

SOURCE: USDA 1988.

After the land area is stratified by use, the strata are further partitioned into areas that vary in size according to production intensity. These sampling units are subdivided again into segments of uniform size. These segments provide the population from which random samples can be drawn. Sampling occurs at a greater rate for segments of interest. Thus, intensely cultivated farmland may have a greater sampling rate than nonagricultural areas. The sampling rate for different land-use segments is determined by the desired precision of the estimates for the commodities of interest.

The actual survey of the segments is conducted by using three reporting methods—closed segment, open segment, and a weighted segment. The closed segment records land use of the resident operator within the closed boundary of the segment. Parcels of land operated by the resident producers outside the segment boundary are not included in the survey. The open segment method records total land use of the resident operator. Thus, land operated by resident operators within the bounded segment, as well as production outside the segment, is recorded in the survey. In the weighted segment, a survey of the total land of resident operators is weighted by the proportion of land within the sample segment boundary.

For the *Hogs and Pigs* report, the sampled land segments are augmented with lists of large producers in each surveyed state. These large operators are sampled at a greater rate, reflecting their importance in determining the hog supply. These operators are not included in the list frame. Survey data for the area frame are collected through personal interviews. This is a relatively expensive enumeration method.

The primary disadvantage of the area frame sample is cost. With the area frame method, all tracts of land need to be identified, partitioned, and segmented by use. After the area frame is established, however, it can be used for a long period because adjustments in land-use patterns are typically slow. The USDA updates three to four states per year in the area frame (USDA 1983). Enumeration by personal interview also adds to the survey cost.

Another disadvantage specific to livestock production is that the area frame does not provide accurate estimates of items that are poorly correlated with land area. The area frame is preferred for obtaining estimates of widely produced agricultural commodities, items that appear frequently within the domain of the segment. Thus, the area frame provides better estimates of major crops, such as corn and wheat, that are widely produced geographically.

Livestock estimates from the area frame sample are less reliable because production is often geographically concentrated and has significant production size variability. In the area frame, it is difficult to adequately represent commodities that appear in less than one segment in five, or are produced on less than one-fifth of the farms (USDA 1983). Also, imprecise boundary definitions can create biases in the sample estimates from the area frame (Jessen 1978). This boundary problem is compounded in the case of livestock, because livestock can move easily from the bounded segment.

Nevertheless, the primary advantage of the area frame method is that it provides a complete frame. All members of the population have a known probability of being included in the sample. Also, the area frame does not go out of date quickly unless land-use patterns and other characteristics of the population change quickly. The area frame does provide an efficient method of obtaining supply estimates for commodities that appear frequently in sampled segments.

List Frame

List frames are used in conjunction with area frame surveys to complement the deficiencies inherent in the area frame survey method. The list frame survey is a less expensive method of data collection. For the hog survey, the list frame is simply an array of producers that may raise hogs within a state. The list frame affords a more efficient means of data collection than the area frame, but is nearly always an incomplete representation of the entire population. Maintaining complete lists of all hog producers within a state is nearly impossible. Also, the list frame may become outdated quickly. Continuous updating and verification of the list frame are required to maintain accurate control data for stratification.

The list of operators is stratified within each state similarly to the area frame, but the stratification is by characteristics of the farm operation, such as type and size of operation. An example list frame stratification is provided in Table 2.5. Stratification categories include 10,000 hog capacity and between 200 and 599 acres of cropland. The stratification is designed to ensure that producers of rare commodities are represented and that large operators in livestock production capacity, cropland acreage, and storage capacity are sampled. Large operators included in the area frame are removed from the list frame to avoid duplication.

Table 2.5. Example of list frame stratification by production characteristic

Stratum	Population	Sample	Sampling Rate
Cropland 1-999 acres	16,227	270	60.10
Cropland 200-599 acres	9,357	210	44.56
Hogs 1-99 head	12,790	250	51.16
Hogs 100-199 head	7,541	210	35.91
Cropland 600-3999 acres	2,585	85	30.41
Hogs 200-399 head	8,657	410	21.11
Hogs 400-599 head	4,016	260	15.45
Hogs 600-999 head	3,199	275	11.63
Hogs 1000-1999 head	1,609	225	7.75
Hogs 2000-3999 head	304	100	3.04
Hogs 4000-9999 head	44	20	2.20
Cropland >4000 acres	15	15	1.00
Hogs >10,000 head	5	5	1.00

SOURCE: USDA 1988

The large operators and producers of more specialized commodities are sampled at higher rates. For example, hog producers with more than 10,000 head capacity are sampled at a greater rate than producers with less hog production capacity. Combining stratification and the greater sampling rates for certain production and size characteristics, the list frame can account for operators that are essentially outliers, but that have a substantial impact on the supplies of the underlying commodities.

The list frame survey contains about 77,800 producers for the December survey, slightly fewer for the June survey, and about 25,700 producers in the March and September surveys (USDA 1988). The survey is conducted principally by telephone interview, with additional enumeration through mailed questionnaires and by personal interviews.

The list frame provides an efficient sampling method of items that appear infrequently within the sampled units or when there exists extensive variability in size of operations. These features are characteristic of livestock enterprises in general and hog operations in particular. Large hog operations can be sampled at a greater rate and thus be incorporated in the survey estimate. Also, sampling from the list frame is relatively inexpensive compared with the area frame sampling. However, the list frame sample is nearly always incomplete, and often it is costly, if not impossible, to maintain a current list of producers and their production characteristics.

Multiple Frame

Multiple frame estimates combine the area and list frame estimates. The multiple frame method has many similar attributes of the list frame; it is an efficient means of deriving estimates for items that are poorly correlated with land area. Again, this feature is particularly useful for livestock supply and inventory estimates. With the multiple frame, however, the incompleteness of the list frame is factored into the estimates by incorporating the area frame.

The multiple frame method requires that every element of the population belong to one of the sampling frames. Because the area frame is complete, this criterion is met. Also, for each selected element, it must be possible to identify whether a selected element is contained in the other sampling frame. Thus, all sample operators from the area frame sample that also are contained in the list frame sample must be identified.

The multiple frame and area frame estimates provide the basis for estimates in the *Hogs and Pigs* report. For the quarterly hog estimates, the multiple frame combines a relatively large list frame with a small area frame. Direct expansion of the list frame represents a large proportion of the variance of the total population. In Figure 2.4, segmentation of the area and multiple frame samples is given. The segment operators, sampled from the area

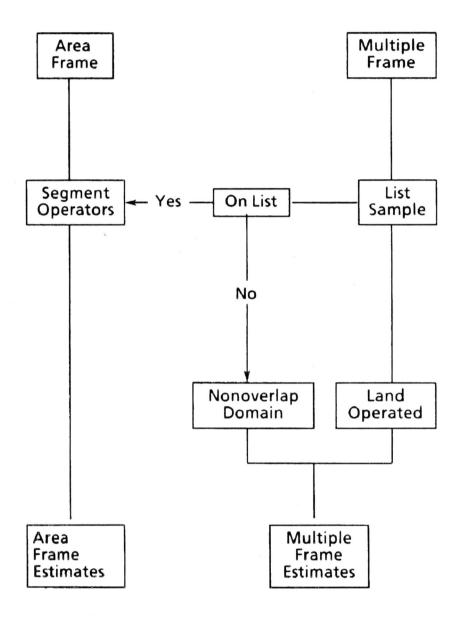

Figure 2.4.   Segmentation of the area and multiple frame
samples (USDA 1988)

frame, are augmented by large producers in the list frame. Again, this ensures that large operators are represented in the sample. The area frame estimate is derived from this augmented sample of segment operators by direct expansion.

The multiple frame estimate is obtained from the list frame sample, and also incorporates large producers from the area frame, which are not included in the sampled segment operators. This latter set of operators is included in the nonoverlap domain. The estimates from the nonoverlap domain and the list sample are combined to form the multiple frame estimate by

$$X = X_a + p * X_{al} + q * X_{al} , \qquad (2.1)$$

where X is the multiple frame estimate, X_a is the estimated total for the portion only included in the area frame, X_{al} is the estimated total for the population included in both frames and computed from the area sample, and X'_{al} is the estimated total for the population included in both frames and computed from the list sample (USDA 1983). The weights, p and q, sum to one, and are inversely proportional to the associated variances with each estimate.

The use of multiple frame and area frame estimates for the *Hogs and Pigs* report is delineated in Table 2.6. For the March and September reports, the ten-state estimate of the hog supply and its disposition is based only on the multiple frame, with no overlapping with the area frame sample. Because the list sample is large relative to the size of the area frame, the list sample provides a reliable estimate of the total population. Again, the list sample is delineated in strata by size of operation and is supplemented by large operators omitted from the sample.

For June and December reports, the multiple frame and area frame sampling techniques, as illustrated in Figure 2.4, are used to derive the estimate. Multiple frame estimates are combined with area frame estimates

Table 2.6. *Hogs and Pigs* report survey methods and coverage

Survey date	Area	Survey Method
December 1	10 quarterly states[a]	Multiple frame and area frame
	13 other states[b]	Multiple frame and area frame
	27 other states[c]	Area frame
March 1	10 quarterly states	Multiple frame
June 1	10 quarterly states	Multiple frame and area frame
	40 other states	Area frame
September 1	10 quarterly states	Multiple frame

SOURCE: USDA 1983, 69.

[a]Georgia, North Carolina, Indiana, Illinois, Missouri, Iowa, Minnesota, Kansas, and Nebraska.

[b]North Dakota, South Dakota, Oklahoma, Texas, Wisconsin, Michigan, Kentucky, Tennessee, Mississippi, Alabama, South Carolina, Virginia, and Pennsylvania.

[c]Alaska estimates are from a nonprobability mail survey, and Hawaii conducts a probability mail survey.

for the ten-state estimate. For the remaining states, either the same procedure is used or only the area frame is used to form the aggregate U.S. estimate.

Estimation Procedures

The estimation procedure combines the efforts of state statistical offices with officials from NASS, an agency of the USDA. The phases of the estimation procedure are depicted in Figure 2.5. The state estimates from the

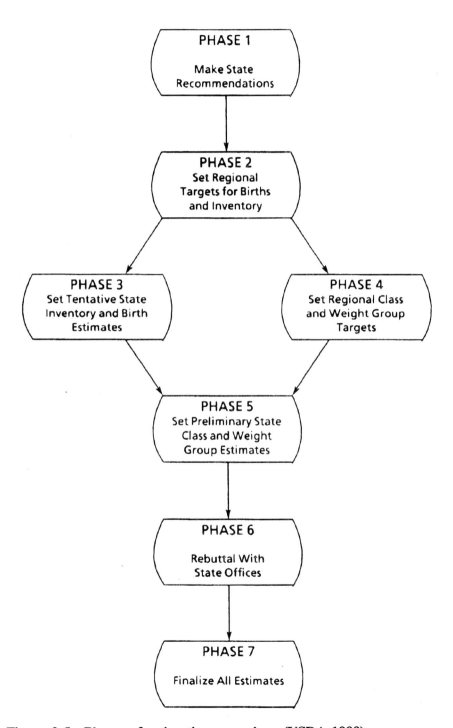

Figure 2.5. Phases of estimation procedure (USDA 1988)

multiple frame and area frame surveys are combined at the national level in the estimation process. At the state level, however, questionnaires are cross-checked to eliminate adding and recording errors and to find possible outliers in the data. The state-level data are further examined for extreme variations and outliers at the national level processing.

Further cross-checks are done at the national level. Comparisons across states are made, and factors such as weather conditions, economic climate, comments from producers, and credit availability also are included in the estimation procedure. Time-series graphs of the previous survey data, percentage-of-total charts for the various supply components, and comparisons with historical relationships are other tools used to develop supply and inventory estimates. Supply and use balance sheets provide further checks to help ensure estimate reliability.

Nonresponse is a factor that must be incorporated into the estimates. Nonresponse can occur because of the inability to locate the individual operator and from operator unwillingness to cooperate. The latter is an option because the survey is voluntary. During the June 1987 survey, the nonresponse for the area frame was about 8 percent, and the nonresponse for the list survey was 15 percent (USDA 1988).

In the *Hogs and Pigs* report, nonresponse is handled in a number of ways, depending on the amount of information available on the operation of the nonrespondent. If there is no information on the operation of the nonrespondent, an average of respondents is used. With partial knowledge, an average of reports indicating hogs on the farm is used, or a zero value is used if it is known that the operator is not currently raising hogs. Also, observation by enumerators and substitute respondents are used in the adjustment for nonrespondents.

The inventory estimates usually are adjusted for nonresponse because information can be gathered on the number of hogs on the farm of the nonrespondent. Estimates of supply flows such as the pig crop, sows farrowing, farrowing intentions, and pig saved per litter are not adjusted because these items are not observable at a given time and require operator participation in the survey.

The bias introduced by nonresponse depends on differences in the characteristics of the observed and unobserved elements and the fraction of the sample observed (Jessen 1978). The bias of nonresponse for estimating y_r, the mean of the sample, is given by

$$\text{bias } (y_r) = (N_{nr} / N) * (Y_r - Y_{nr}), \tag{2.2}$$

where N_{nr} is the number of nonrespondents in the sample, N is the sample size, and $(Y_r - Y_{nr})$ is the difference between the true observed and the nonobserved mean values (Jessen 1978, 460). Thus, as the number of nonrespondents increases, or as differences between respondents and nonrespondents increase, the bias in the estimates increases.

1987 Survey Changes

Before 1987, the quarterly *Hogs and Pigs* reports estimated the March 1, June 1, September 1, and December 1 hog supplies and inventories and were released between the 20th and 23rd of the respective months at 3:00 p.m., Eastern daylight time. Because the survey process requires about one month from the start of data collection to release, the surveys frequently were completed before the first day of the release month. Thus, surveys did not reflect supply conditions at the first of the month, as the report indicated. This was especially problematic for pig crop estimates since operators could not readily predict their future pig crops.

Beginning in 1987, the survey data were collected the first of the release month. Thus, producers could more accurately count the number of hogs, their weights by grouping, and, more important, account for the pig crop born up to the first of the month. The move forward in the start data collection period moves the release date to the end of the month, but still at 3:00 p.m., Eastern daylight time.

Estimate Reliability

The *Hogs and Pigs* reports provide only an estimate of the total hog population and inventories. The entire population is not included in the survey, so sampling error exists and, as with any survey, nonsampling error is present. Nonsampling error reflects omission, duplication, self-selection, and other operative errors. Sampling error is reduced by the series of cross-checks completed before the report is released.

The relative standard errors (standard error of the estimate divided by the estimate) for the December 1987 survey estimate of the hog inventory is about 1.8 percent at the aggregate U.S. level (USDA 1988). The relative standard errors for the breeding herd and market hog estimates for the entire United States are 1.9 and 1.8 percent, respectively. The standard errors for the market hog weight groups are somewhat larger, and range from 2.0 to 2.8 percent, depending on the weight group (USDA 1988). Sampling variability of the supply flows is greater because surveyed operators must remember levels for the past three months. Also, errors in the supply flows reflect underlying errors in the inventory numbers. Hence, estimates of farrowings and the pig crop are less reliable in sampling variability. Of course, the farrowing intentions are the least precisely estimated. Producers may change farrowing plans as more information becomes available about market conditions and as weather conditions change, or as a response to government policy changes.

The relative standard errors for the ten-state estimates in March 1 and September 1 are slightly smaller. The relative standard error for the breeding and market hog inventory categories in the September 1988 report are 1.6 and 1.5 percent, respectively. Also, the sampling variability for the market weight groups is smaller, as measured by its relative standard error, than the aggregate U.S. estimates. For the ten-state estimates, the relative standard errors for the market weight categories range from 1.8 to 2.6 percent.

Revisions of the Estimates

The estimates of the hog inventory are revised when subsequent data suggest their reliabilities are above tolerable levels. Revisions provide a more accurate basis for comparison for future reports. The reliability of the estimates is checked primarily by comparing it with subsequent commercial slaughter levels. Slaughter data are considered more reliable and are collected at the slaughter plant. Late survey questionnaires and errors in past surveys also are factored into the revision process. The U.S. Census of Agriculture, undertaken every five years, provides additional information used to finalize the estimates.

Nevertheless, the primary means of verifying the estimates is through comparing the reported supply levels to future slaughter statistics. If discrepancies appear between the inventory estimates and subsequent slaughter, revisions are made. Revisions are published in the next *Hogs and Pigs* report. Also, the December survey, which is based on the most extensive survey, can provide further evidence that the estimates require revisions. The estimates are subject to further review after the U.S. Census of Agriculture data become available. Thus, final estimates for the hog supply are published about every five years.

In Figures 2.6 through 2.9, the percentage changes between the first and final estimates are provided for selected supply and inventory categories in the June and December *Hogs and Pigs* report. Figures 2.6 and 2.7 give the revisions in the U.S. farrowing and pig crop estimates. Revisions of the estimates for total market hogs and hogs kept for breeding are presented in Figures 2.8 and 2.9, respectively.

In Figures 2.10 through 2.13, the percentage changes between the first and final quarterly estimates are provided for the ten-state region surveyed.

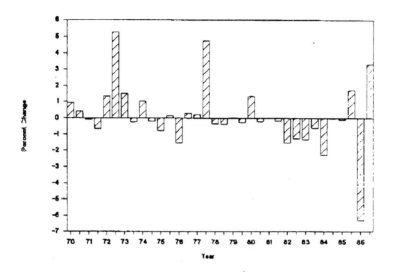

Figure 2.6.   Percentage change between initial and final
estimates: U.S. sows farrowing, 1970-86

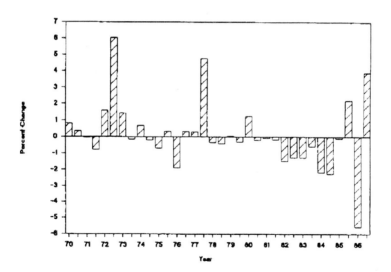

Figure 2.7.   Percentage change between initial and final
estimates: U.S. pig crop, 1970-86

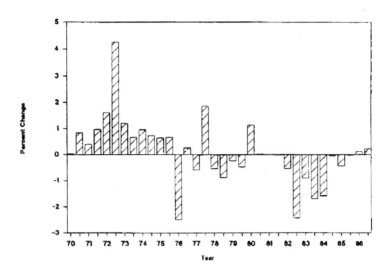

Figure 2.8. **Percentage change between initial and final estimates: U.S. market hogs, 1970-86**

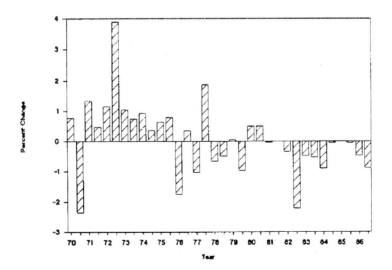

Figure 2.9. **Percentage change between initial and final estimates: U.S. hogs kept for breeding, 1970-86**

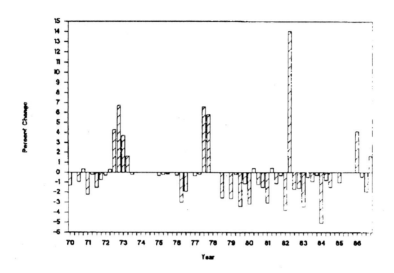

Figure 2.10.   Percentage change between initial and final
estimates: Ten-state sows farrowing, 1970-86

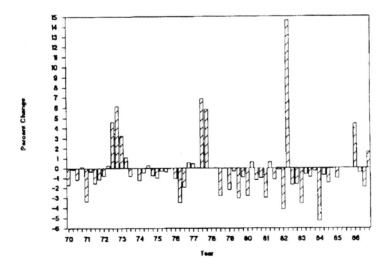

Figure 2.11.   Percentage change between initial and final
estimates: Ten-state pig crop, 1970-86

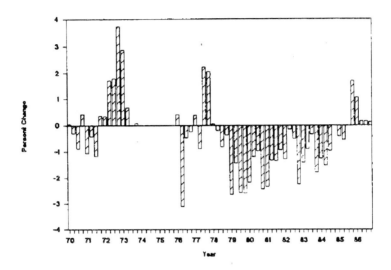

Figure 2.12. Percentage change between initial and final estimates: Ten-state market hogs, 1970-86

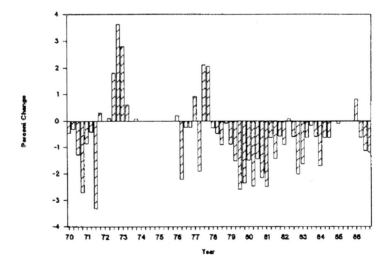

Figure 2.13. Percentage change between initial and final estimates: Ten-state hogs kept for breeding, 1970-86

The revisions in the level of farrowings and the pig crop are presented in Figures 2.10 and 2.11, respectively. Figures 2.12 and 2.13 present revisions of the estimates of market hogs and hogs kept for breeding, respectively.

The estimates since 1982 were subject to revision until the 1987 U.S. Census of Agriculture was finalized. The 1982 second quarter estimates of hogs farrowing and pig crop are both more than 14 percent. This substantial revision was made, in part, because of the redefinition of the ten-state region in that year.

Most of the revisions are less than 2 percent. However, some initial estimates are beyond the confidence intervals provided previously. For example, the December 1972 U.S. farrowing and pig crop estimates were revised upward by 5.3 and 6.1 percent, respectively (Figures 2.6 and 2.7). For that same report, revisions of the estimates of market hogs and hogs kept for breeding were revised upward by 3.9 and 4.3 percent, respectively (Figures 2.8 and 2.9). Similar upward revisions were made in the December 1977 estimates for the same categories.

Downward revisions of 6.3 and 5.6 percent were made in the June 1986 estimates of U.S. farrowings and the pig crop, respectively. However, essentially no revisions were made in the estimates of market hogs and hogs kept for breeding in that same report. The estimate revisions for the ten-state region follow the revisions in the U.S. estimates, but the ten-state revisions are larger and more frequently exceed the 3 percent level.

Summary

The revisions illustrate the level of estimate error in the reports. Overall, the *Hogs and Pigs* report provides a useful account of hog supply and inventory movements. The survey methodology has advanced, improving reliability of the estimates since the USDA began collecting data. The survey methodology has adapted to changes in the structure of the industry. However, as with any voluntary survey, there is some nonresponse. The biases of nonrespondents have increased as the structure of the hog industry has shifted to more variability in production characteristics. The likelihood of respondents and nonrespondents being similar is less.

The survey and sampling methods have incorporated probability sampling techniques such as multiple frame sampling to reduce estimate error. However, the USDA has not advanced its data evaluation techniques. The estimates still rely primarily on survey results and the judgment of NASS officials in identifying and handling outliers, in adjusting for nonrespondents, and in deriving the initial estimates using other processes.

Improved market information is being demanded because investment in the hog industry has increased and the abilities of producers to process timely market information has grown. Improved estimates of hog supply and inventory changes could be developed by expanding survey coverage, but expanding this coverage is costly, and the biases of nonrespondents remain. The information base of the estimates also could be expanded to include hog supply and inventory estimates based on econometric models. These econometric models incorporate biological constraints and price expectations based on rational and futures market expectations. A composite forecast of the survey and econometric models would increase the accuracy of the initial estimates without additional sampling.

Chapter 3

The Generation
of Price Expectations

Expectations, as defined by Muth (1961, 4), are "informed predictors of future events." Individuals form expectations about future events because decisions made in the present create opportunities and limits on future production, consumption, and wealth. Production lags, both technological and biological, and investment requirements require individuals to adopt a planning horizon for many production and consumption decisions. Individuals incorporate past economic relationships and knowledge of current economic conditions in their decision processes, as well as their expectations of future economic activity. Thus, future expectations are based on past experience and perceptions about how the structure of the associated economic system functions.

In econometric models, the mechanisms used to derive individuals' expectations are critical because they are designed to represent the processes by which decisions determining consumption, production, and investment are developed. Expectations must be depicted precisely to provide an accurate representation of the individuals' behavior (Chavas and Johnson 1982). However, accurately expressing expectations is difficult because they are unobservable. Various hypotheses have been advanced that yield estimable structures for representing expectations of the behavior of agents in econometric systems.

41

The dominant form of the price expectation hypothesis used in econometric models involves a form of distributed lag in past prices. That is, predictions of future prices are based on the historical series of past prices. Past prices are readily observable and provide an easily implemented method of generating price expectations.

Futures market prices are another easily applied means of representing price expectations. Futures market prices reflect the traders' average price expectations of spot prices that will prevail at the contract's maturity (Telser 1958; 1967). Differences in the future and current spot prices incorporate foregone costs of storing the commodity (Kaldor 1939; Working 1948; Brennan 1958; Telser 1958) and possibly an expected risk premium (Cootner 1960; Dusak 1973; Breeden 1980). If market information is freely disseminated and markets are efficient (Fama 1970), there is no reason for traders and other market participants to have different expectations (Gardner 1976). The rational expectations hypothesis (REH), as developed by Muth (1961), expands the basis for characterizing the information set for the individual beyond extrapolations of historical price series. With rational expectations, the information set is specifically defined as the system of stochastic equations that form a representation of the econometric system. Thus, in an econometric modeling context, individuals are assumed to know the model structure and use it as the basis for their expectations about future events.

This chapter reviews the alternative expectation hypotheses. The alternative representations are used here to expand the information set on which the hog supply and inventory estimates are based. The pork industry models that incorporate rational and futures price expectations yield predictions of hog supply and inventory categories that do not solely depend on survey results. Given desirable properties of composite forecasts, improved estimates of the supply and inventory categories contained in the *Hogs and Pigs* report can be obtained.

Distributive Lag Expectations

In his seminal paper on the cobweb model, Ezekiel (1938) used naive expectations in his exposition on reasons for oscillations in commodity markets. Naive expectations are the simplest form of the distributed lag expectation formulation. The expected price, y_t^*, is defined as the past period's price, y_{t-1},

$$y_t^* = y_{t-1} .$$
(3.1)

Initially, this form of price expectation was hypothesized to conform with discrete production structures. Production decisions occurred in the previous period because of a one-period production lag, and thus producers used the current price as the expected price. Price is then determined by predetermined available supply, coupled with the current period's demand. Naive expectations are used frequently in continuous production structures, such as livestock, because of their ease of implementation and because of their success in capturing the behavior of individuals in econometric models of the agricultural sector. For example, Coase and Fowler (1937) examined the pig market cycle in Great Britain by using naive expectations with the cobweb model detailed by Ezekiel.

Metzler (1941) and Goodwin (1947) expanded individuals' information sets. The expected price was determined by the price of the previous period, plus a certain fraction of the difference of the past two period prices,

$$y_t^* = y_{t-1} = \alpha(y_{t-1} - y_{t-2}) \ . \tag{3.2}$$

Thus, in addition to the past period's price, individuals in this model incorporate information on price trends.

Another prevalent form of price expectations used in commodity market models is expectations derived from further extrapolations of past prices. Adaptive expectations, as developed by Cagan (1956), popularized by Nerlove (1956; 1958) and extensively reviewed by Askari and Cummings (1977), are an extrapolative form of price expectation. Adaptive expectations are based on the premise that agents revise their expectations by examining the past period's forecast errors. Algebraically, this form of price expectation is

$$y_{t+1}^* - y_t^* = \alpha(y_t - y_t^*) \ , \tag{3.3}$$

where the revision of price expectations, $y_{t+1}^* - y_t^*$, is determined by a fraction of the previous period forecast error. By simple manipulation, adaptive price expectations reduce to expectations based on the series of past prices with geometrically declining weights:

$$y_{t+1}^* = \alpha \sum_{j=0}^{\infty} (1 - \alpha)^j \, y_{t-j} \ . \tag{3.4}$$

Other parsimonious representations of distributed lag structures with more flexibility in the lag structure include those by Almon (1965) and Jorgenson (1966). Almon reduced the parameters to estimate by approximating the lag structure with a polynomial of low degree. Jorgenson developed a rational lag structure that represents the distributed lags as a ratio of two polynomials.

The main criticism of distributed lag price expectations is that individuals waste information. With any form of distributed lag expectations, individuals can make systematic and persistent forecasting errors in their decision making processes. Another criticism of distributed lag expectations is that they are ad hoc (Nerlove 1979). The geometric decline in the weights attached to past prices is only a restriction imposed to circumvent estimation difficulties and is not derived from an optimization process. This criticism also holds true for other, more general, distributed lag formulations such as those developed by Almon (1965) and Jorgenson (1966).

Futures Market Price Expectations

Futures market prices are another mechanism to represent expectations in models developed by Working (1942) and in supply response models implemented by Gardner (1976), Subotnik and Houck (1982), and by Chavas, Pope, and Kao (1983). Gardner justified expectations based on futures market prices by using the earlier work of Telser (1967, 174), who stated that ". . . futures prices can be considered as an unbiased prediction of subsequent spot prices." However, this conjecture is controversial. There is no agreement on whether futures prices have the power to predict future spot prices (Fama and French 1987).

In part, the controversy reflects the roles of futures markets in the functioning of markets. Futures markets help to shift price risk from hedgers to speculators (Keynes 1930; Hicks 1939). Hedgers pay a risk premium to speculators for accepting the price risk. Thus, a risk premium, if it exists, may bias the predictive accuracy of futures market prices as price predictors (Cootner 1960; Dusak 1973; Breeden 1980). Also, as Peck (1976) notes, futures markets, by shifting the risk of holding inventories, have an allocative role. For storable commodities, the difference between the current spot price and the futures prices is the net marginal cost of storage (Kaldor 1939; Working 1948; Brennan 1958; Telser 1958). Futures markets also provide price discovery and market information mechanisms (Hieronymus 1971; Peck 1976).

Controversy about the use of futures market prices as price predictors also encompasses the question of futures market efficiency. Futures market efficiency, as defined by Fama (1970), is the incorporation of all available and relevant information into the futures market quotation. Working (1942) initially questioned the futures markets as price predictors. More recently, Grossman and Stiglitz (1980) and Bray (1981) provided theoretical arguments that futures markets cannot reasonably incorporate all available information.

Setting aside the controversies that exist, futures market prices as representations of price expectations still do incorporate the participants' expectations of level spot prices at contract maturity and do provide additional market information. Thus, including futures market price expectations in the pork industry model provides an alternative information set on which to base forecasts of the supply and inventory categories.

Futures Market Prices as Price Predictors

The empirical tests of futures market efficiency are many and often contradictory because of the mixed sets of commodities, time periods of analysis, and efficiency tests (Tomek and Robinson 1981). The empirical tests are based primarily on finding a better forecasting mechanism, typically defined by minimum mean square error (MSE), than the futures market's prediction. If such a mechanism cannot be found, the futures market is judged efficient. Rausser and Carter (1983) contend that the MSE-based test only provides proof of market inefficiency. They contend that the analysis must be extended to show that the superior forecasting method can generate greater risk-adjusted profits when cost of usage is included.

In the set of nonstorable commodities (livestock, potatoes), results have been mixed with respect to futures markets' ability to predict prices. In general, futures markets perform well for short-term predictions of spot prices, but are inferior to other forecasting mechanisms for longer time horizons. Nevertheless, the results seem sensitive to the periodicity of the data, the sample period, and the methodology employed.

The results of Tomek and Gray (1970) suggest that spring futures market quotations for Maine potatoes do not provide a price forecast of subsequent spot prices during fall harvest. More germane, Just and Rausser (1981, 203) found that "futures prices perform quite well as forecasters for . . . hogs," but that forecasts from econometric models were superior beyond the one-quarter time horizon. Leuthold and Hartman (1979; 1981) found that econometric models provide more accurate forecasts of hog prices. These results were confirmed and extended by Leuthold et al. (1987). Hudson et al. (1985, 61) found that the live hog futures market is efficient, but is an "information-starved" market. Recently, Fama and French (1987) found that futures market prices for hogs had reliable forecasting power and found a connection between basis variability and predictive power.

Results were similar for other livestock markets. Leuthold (1974, 379) suggested that, for cattle "from about 15 to 36 weeks prior to delivery, one can expect a better estimate of the futures cash price of cattle by looking at the present cash price than by studying the futures price itself." Stein (1981, 228) found that, for live cattle, "producers received misleading signals from futures prices" for more than four months from maturity. Wilkinson (1985) found that an econometric model provided better forecasts of live cattle price than did futures market prices.

Therefore, using futures markets as the basis for price expectations is not always a clear substitute for expectations based on even simple forecasting rules. Futures market prices do carry information, but, as Working (1942, 49) contends, are not forecasts "in the sense in which one speaks of the price forecasts of a market analyst." This has been confirmed by the studies just mentioned in which time-series methods and econometric models based on distributive lag price expectation mechanisms outperformed futures prices as forecasts of future spot prices. Futures market prices may not give an unbiased estimate of future spot prices; nevertheless, they can expand the information on which the estimates of the hog supply and inventory categories are based.

Rational Expectations

Rational expectations assume that individuals are aware of their surroundings and use this information to predict future events. This hypothesis implies that individuals have the ability to process and use information efficiently in their decision making. The hypothesis is based on the assumption that "information is scarce, and economic systems generally do not waste it" (Muth 1961, 5).

The REH extends the basis of expectations beyond extrapolations of past behavior and assumes that individuals have ability to incorporate new information about both the system's structure and changes in policy. Earlier, Heady and Kaldor (1954) found evidence to support Muth's assertions. They found that, for a sample of Iowa farmers, a common price-predicting mechanism was to use the current price and adjust for "the expected effects of important supply-and-demand forces" (Heady and Kaldor 1954, 35). Forecast revisions were made because of "unforseen events and new information," and, in revising their forecast, the farmers tended "to reappraise the price outlook to arrive at a new set of estimates" more aligned with their "current information" (Heady and Kaldor 1954, 44).

Muth (1961, 4) asserted that individuals' predictions "are essentially the same as the predictions of the relevant economic theory," and thus "the way expectations are formed depends specifically on the structure of the relevant system describing the economy." More specifically, "expectations of firms (or, more generally, the subjective probability distribution of outcomes) tend to be distributed . . . about the prediction of the theory. . . ." (Muth 1961, 5).

Thus, the unobservable expectations of individuals are about the same as the mathematical conditional expectation implied by the structural model. The structural model defines the information set of individuals in the modeled economic system. This has implications of econometric models that represent systems of economic activity. It forces a consistency between the structural representation of the stochastic system and the expectation mechanism used by the system's participants in their decision making.

However, the REH does not preclude differing expectations among individuals, nor does it presume that all individuals use the same information set in forming their expectations. Rather, the REH asserts that their average expectation of future events can be condensed in the structural model. The structural model, in turn, gives an average of individuals' consensus of the system's structure and its underlying behavior.

To clarify, a simple example of implementing rational expectations in a linear model is given. The example was initially presented by Wallis (1980), and for convenience uses his same notation.

REH in a Simple Linear Model

Consider the two-equation linear model in matrix form as provided in (3.5). The model has two observable endogenous variables, y_{1t} and y_{2t}, an observable exogenous variable x_t, and the expected value of one of the endogenous variables y_{1t}^* formed in period t - 1. This latter unobservable variable, y_{1t}^*, the expected value of y_{1t}, is defined as the expectation implied by the model conditional on the available information, Φ_{t-1}, i.e., $y_{1t}^* = E(y_{1t} \mid \Phi_{t-1})$. The expected value of y_{1t} is included as an input into the determination of y_{2t}. The disturbances for the first and second equation are u_{1t} and u_{2t}, respectively. The disturbances in the structural model are assumed to have a zero mean and are serially uncorrelated. The parameters of the model are assumed to be known with certainty by the agents in the modeled system. In essence, the individuals know the structure of the system, and use it as the basis for their expectations.

$$\begin{bmatrix} 1 & \beta_{12} \\ \beta_{21} & 1 \end{bmatrix} \begin{bmatrix} y_{1t} \\ y_{2t} \end{bmatrix} + \begin{bmatrix} \alpha \\ \gamma \end{bmatrix} \begin{bmatrix} y_{1t}^* \\ x_t \end{bmatrix} = \begin{bmatrix} u_{1t} \\ u_{2t} \end{bmatrix}. \tag{3.5}$$

The reduced form of (3.5) can be obtained, provided the usual matrix rank conditions hold, as given in (3.6). Because the reduced form contains an unobservable component, y_{1t}^* the expected value of y_t is operationally infeasible. The form of expectations for y_{1t}^* must be assumed before the model can be implemented.

$$\begin{bmatrix} y_{1t} \\ y_{2t} \end{bmatrix} = \frac{1}{1-\beta_{12}\beta_{21}} \begin{bmatrix} -\alpha & \gamma\beta_{12} \\ \alpha\beta_{21} & -\gamma \end{bmatrix} \begin{bmatrix} y_{1t}^* \\ x_t \end{bmatrix} + \begin{bmatrix} v_{1t} \\ v_{2t} \end{bmatrix}. \tag{3.6}$$

The transformed disturbances in the reduced form, v_{1t} and v_{2t}, still have a zero mean and are serially uncorrelated because they are linear combinations of disturbances with these same properties. The reduced form (3.6) can be written more compactly as

$$\begin{bmatrix} y_{1t} \\ y_{2t} \end{bmatrix} = \begin{bmatrix} \pi_{11} \pi_{12} \\ \pi_{21} \pi_{22} \end{bmatrix} \begin{bmatrix} y_{1t}^{*} \\ x_t \end{bmatrix} + \begin{bmatrix} v_{1t} \\ v_{2t} \end{bmatrix} , \qquad (3.7)$$

where the values of π_{ij} can be inferred.

The REH assumes that individuals' expectations are consistent with the model's structure. The rational expectation of y_t^{*} is the mathematical conditional expectation implied by the model. From (3.7), the rational expectation of y_t^{*} can be derived. Again, this manipulation assumes the matrix $1 - \pi_{11}$ is of full rank. The expectation is conditional on the available information in period t - 1. Thus, if the disturbances have zero mean and are serially uncorrelated, the rational expectation of y_t is

$$y_{1t}^{*} = E(y_{1t} \mid \phi_{t-1}) = (1 - \pi_{11})^{-1} \pi_{12} \ E(x_t \mid \phi_{t-1}) . \qquad (3.8)$$

The rational expectation of y_t^{*} depends on the conditional expectation of the exogenous variable, x_t. Typically, it is assumed that individuals have limited information about the structure that generates x_t and consequently use an optimal extrapolative predictor. For this example, the optimal predictor of x_t is defined as

$$x_t = \phi \ x_{t-1} + \epsilon_t , \qquad (3.9)$$

where ϵ_t is assumed to have a zero mean and be serially uncorrelated. Given the information available in period t - 1, the predictor of x_t is defined as the conditional expectation

$$\hat{x}_t = E(x_t \mid \phi_{t-1}) = \phi \ x_{t-1} . \qquad (3.10)$$

After substituting (3.10) into (3.8), the rational expectation of y_{1t}^{*} becomes

$$y_{1t}^{*} = (1 - \pi_{11})^{-1}\pi_{12} \; \phi \; x_{t-1} \; . \tag{3.11}$$

The rational expectation of y_{1t}^{*} can then be substituted in the unobservable reduced form. This creates an estimable two-equation system that Wallis calls the "final form." Equations 3.12 and 3.13 give the final form of the system, which relates an observable endogenous variable to the exogenous variables. This form is fully estimable and contains expectations consistent with the structure of the model.

$$y_{2t} = (1 - \pi_{11})^{-1}\pi_{11} \; \pi_{12} \; \phi \; x_{t-1} + \pi_{12}x_{t} + v_{1t} \; , \tag{3.12}$$

and

$$y_{2t} = (1 - \pi_{11})^{-1}\pi_{21} \; \pi_{12} \; \phi \; x_{t-1} + \pi_{22}x_{t} + v_{2t} \; . \tag{3.13}$$

The rational expectation model has several desirable properties that reflect the optimal use of information by the system's agents. By subtracting (3.11) from (3.12), the rational expectation forecast error is defined as

$$y_{1t} - y_{1t}^{*} = \pi_{12}(x_{t} - \hat{x}_{t}) + v_{1t} \tag{3.14}$$

The rational expectation forecast error (3.14) can be written as a linear combination of two serially uncorrelated random disturbances that have mean zero:

$$y_{1t} - y_{1t}^{*} = \pi_{12} \; \epsilon_{t} + v_{1t} \; . \tag{3.15}$$

Thus, the rational expectation forecast error is a linear combination of the error associated with projecting the exogenous variables, ϵ_{t}, and the error associated with random shocks to the system, v_{1t}. Given the assumed properties of the disturbances, the rational expectation forecast error is serially uncorrelated and has a zero mean (Wallis 1980). In the rational expectations model, individuals use all available information, as defined by the model structure, to form their expectations optimally. Also, the rational expectation forecast error has a lower error variance than the optimal extrapolative predictor (Nelson 1975; Wallis 1980).

The REH directly affects the structure of econometric models, because it is a function of predictions of exogenous variables. This implies that the exogenous variables contained in the model must be projected with some degree of reliability to be useful in the formation of expectations of the endogenous variables (Feldstein 1971; Chavas and Johnson 1982). This requirement limits the conditioning variables in the models to the set that can be projected with some accuracy. Also, this limits the length of the relevant planning horizon. The planning horizon is limited by the time span in which the conditioning variables still contain informational, and hence predictive, content.

Estimation and Applications of the REH in Linear Models

The REH primarily has been incorporated in linear models similar to the one described by Wallis (1980b). The rational expectation model of the pork industry used in this study, however, is posited as nonlinear in its variables. This requires the use of an estimation procedure for nonlinear rational expectation models, developed by Fair and Taylor (1983). Details on their estimation method are presented in Chapter 5. The estimation techniques for linear rational expectation models are presented to provide background and to facilitate discussion.

Substituting the rational expectation of price (3.11) in the structural model (3.5), creates a two-equation system that is nonlinear in its structural parameters. Even with the simple linear model, and particularly with more complex REH models, full implementation of the REH requires appropriate nonlinear estimators. In the simplified example, cross-equation restrictions are not required (the model is just-identified and has only a single lag entering the prediction of the exogenous variable x_t). In more complex linear models, however, the REH implies a set of highly nonlinear cross-equation restrictions.

The highly nonlinear cross-equation restrictions combined with the nonlinear structure imply the use of a full information estimator. This class of estimators contains full information in the essence structural disturbances, and nonlinear restrictions of the model are incorporated in the estimation. Wallis (1980) and Begg (1982) suggest the full-information, maximum likelihood estimator (FIML) and three-stage least squares (3SLS), respectively. The properties and estimation methods required for these estimators are described in Fomby et al. (1984).

Due to the nonlinearities, implementing the REH in linear models of the agriculture sector with full information estimation techniques necessitates simple model structures. Nevertheless, the agricultural supply estimates have been quite successful and are similar to previous adaptive expectation results.

Some selected examples of implementation of the REH with full information estimators include Eckstein (1984) and Tegene et al. (1988). They examined land allocation and supply response for Egypt and Iowa, respectively. Goodwin and Sheffrin (1982) and Phillip (1986) tested the rational expectation hypothesis in the U.S. broiler industry. Shonkwiler and Emerson (1982) estimated a three-equation model of the Florida tomato industry with the FIML estimator. Aradhyula and Johnson (1987) examined the appropriate production lag structure in competing models of the U.S. beef sector. They used nonlinear 3SLS to incorporate the cross-equation restrictions in estimation.

Full information estimators can be burdensome to compute. McCallum (1976), Wallis (1980), and Wickens (1982) have suggested the use of alternative limited information estimators. This class of estimators is termed *limited information* because single equation methods are used and cross-equation restrictions implied by the REH are ignored. This latter trait of limited information estimators prohibits testing the validity of the REH within the modeled structure. However, limited information estimates do provide efficient and consistent estimates of structural parameters (Wickens 1982). The reasoning for limited information estimation techniques can be seen in (3.11). The prediction of the exogenous variable, $\hat{x}_t$, is simply an instrument for the expectation of y_t^*, given the implied structure of the model. Other instruments are available to provide consistent estimates of the structural parameters. Current and lagged values of x_t and lagged values of y_t can be used as appropriate instruments (McCallum 1976; Wallis 1980). Other proxies for the expectation variable include the realized or observed variables. As McCallum (1976) and Wickens (1982) demonstrate, this is an error in the variables method that also provides consistent estimates of the structural parameters. These limited information methods are discussed in detail in Fomby et al. (1985).

Econometric models of the agricultural sector that use the REH with limited information estimation methods still retain the fairly simplified supply and demand structures. For example, Huntzinger (1979) constructed instruments that were lagged values of the price in his model of the U.S. broiler industry. Zanias (1987) estimated a model of tobacco export demand by using the instrumental variable methods proposed by McCallum (1976).

Summary

Each of the alternative expectation mechanisms described incorporates different sets of information to depict the unobservable anticipations and decision processes of individuals. The alternative information sets, in turn, influence the behavior of the system or equations through consumption, production, and investment decisions of individuals. With naive and extrapolative expectations, the information set is limited to past prices. Expectations based on future market prices embody the anticipations of futures market participants. The REH assumes that the information set of individuals is the system of equations. This system represents the known characteristics and processes that govern the structure and behavior of the system.

These expectation mechanisms can provide alternative information sets on which the econometric models of the pork sector rely. Thus, the predictions from the alternative econometric models would contain different market anticipations of individuals. In this way the information set on which the estimates in the *Hogs and Pigs* report are based can be expanded without additional survey coverage.

Chapter 4

The Structure
of the Pork Industry Models

In this chapter, the structures of the quarterly pork industry models are developed and the modeling approach is described. The econometric models of the pork sector incorporate rational and future market expectations.

Review of Econometric Models of the Livestock Sector

Econometric models of livestock have advanced slowly. Most specifications still have relatively simple supply structures that use distributed lags in input and output prices, time lags, and partial adjustment production mechanisms as conditioning variables. Seasonality, an important feature of the livestock sector, is handled with dummy variables. The continued use of this specification reflects, in part, the regularities in the livestock production process, ease in implementation and estimation, and the relative success in capturing industry behavior. Demand specifications are predominantly simple linear structures that do not presuppose adherence to the theory of consumer behavior.

Identifying the underlying reasons for the cyclical nature of pork production and prices provided the initial impetus for modeling the pork economy. The cyclical nature of pork production was initially explained as self-generating by the so-called Cobweb theorem (Coase and Fowler 1937; Dean and Heady 1958; Harlow 1960). Early econometric analyses (Foote

57

1953; Maki 1962) of the livestock feed economy attempted to quantify cyclical price production relationships. These analyses identified the biological sequences inherent in the livestock production process as underlying factors generating the cycles.

The biological sequence in the pork production process remains the benchmark for specifying subsequent econometric models of the pork economy. This is reflected in the recursive supply structure advanced by Harlow (1962) that continues to be essentially replicated in other econometric models. The supply of pork is governed by the number of sows farrowing, which is dependent on past hog and feed prices. Sows farrowing determines hog slaughter, which in turn determines pork production. In this general supply structure, a single inventory relation is specified as a partial adjustment relation, which in turn governs subsequent slaughter. Modern analyses using this structure include Freebairn and Rausser (1975) and Stillman (1985).

Often the supply structure first used by Harlow (1962) is augmented by intermediate steps between farrowing and subsequent slaughter, with equations that represent the pig crop and the levels of market hogs on feed. Also, additional equations are added to represent movements in the size of the breeding inventory. These specifications remain tied to the biological timetable for pork production and include forms of distributed lags in input and output prices. Examples of extended supply structure include Maki, Liu, and Mates (1962), Arzac and Wilkinson (1979), Brandt et al. (1985), Holt and Johnson (1988), and Skold and Holt (1988).

Another set of livestock models, which contain a fairly disaggregate depiction of the supply process, incorporate restrictions implied by the biological process of production. In these models, known biological relationships and seasonality inherent in the production process are incorporated in the behavioral equations. Thus, the biological sequence of production is used to provide more information than defining the lag length of conditioning variables in the supply component.

Incorporating biological restrictions in the supply structure was first done by Johnson and MacAulay (1982) in a quarterly beef model. Historical biological relationships were used to obtain restrictions on the parameter estimates within the supply structure. This approach has been used in livestock models for beef (Okyere 1982; Okyere and Johnson 1987; Grundmeier et al. 1988), poultry (Chavas and Johnson 1982), and pork (Blanton 1983; Oleson 1987; Skold, Grundmeier, and Johnson 1988). Similar approaches include incorporating biological restrictions with the functional form of the supply components. Chavas and Klemme (1986) do so in their analysis of investment behavior of the U.S. dairy industry.

The economic variables included in the inventory specifications have been conditioned on variables beyond distributed lags of input and output prices. Measures of relative profitability in competing enterprises, usually beef production, have been included to reflect the opportunity cost of production. MacAulay (1978) included a beef feeding margin in his pork supply equation. Harlow (1962), Freebairn, and Rausser (1975), Arzac and Wilkinson (1979), and others included producer prices of cattle in their sow inventory equations. MacAulay also included a grain stocks variable to represent feed availabilities.

In many livestock models, the demand equations are estimated in price-dependent form, with per capita meat quantities and income as explanatory variables (e.g., Harlow 1962; Heien 1975, 1977). Fox (1953) suggested this specification because, in the short term, livestock production is essentially fixed. Thus, estimation can proceed with ordinary least squares (OLS). The price-dependent form has not always been followed (Freebairn and Rausser 1975; Arzac and Wilkinson 1979). Nevertheless, in general, the theory of consumer behavior has not been applied in models of the agricultural economy. The standard forms remain linear in the variables and ad hoc in nature (Tomek and Robinson 1977).

The Biological Nature of Hog Production

Even with advances in technology, better nutritional practices, and improved herd management, the biological nature of hog production remains essentially unchanged. The biological process of hog production constrains the response of producers to economic variables. The time lags inherent in the breeding, gestation, and finishing phases of production provide the basis for specifying the supply structure of the model. The biological timetable of hog production is presented in Figure 4.1. The diagram demonstrates the relatively long production response when producers decide to expand their production capabilities. The successive stages of production, delineated in Figure 4.1, are discussed in turn.

After sows and gilts are bred, the gestation period is nearly four months. Sows are hogs that have given birth, and gilts are unbred female hogs. Between 1970 and 1986, sows produced an average of 7.4 pigs per litter (USDA 1977a; 1980; 1983; 1988). Litter size has increased with gains in productivity of the breeding herd. Litter size is typically not dependent on enterprise size, but geographical location of production, along with the age distribution of the breeding herd, does have an influence (Van Arsdall and Nelson 1984). Random factors such as weather and disease also are important influences on the number of pigs saved per litter.

Pigs are weaned about four to six weeks after birth. Larger operations typically wean at younger ages, 3.5 to four weeks of age, whereas smaller enterprises with more seasonal production wean at older ages (Van Arsdall and Nelson 1984). Sows can reenter the production cycle more quickly with younger weaning ages, and thus more pigs can be produced per sow. After

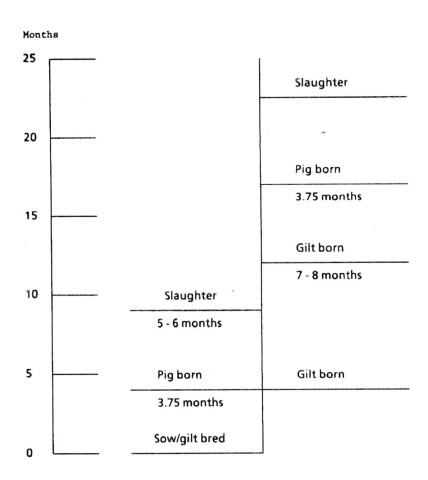

Figure 4.1. Biological timetable of pork production

pigs are weaned, it takes a 40- to 50-pound pig four to five months to reach slaughter weight. Slaughter weights for barrows and gilts are usually 220 to 240 pounds. Barrows are castrated male hogs that are fed for slaughter. The time between breeding of the sow or gilt and slaughter of her offspring is about ten months.

When producers choose to expand their production capabilities by expanding their breeding herd, the biological sequence is essentially the same. Instead of feeding gilts to slaughter weights, however, producers retain these unbred female hogs for breeding. Gilts are bred seven to eight months after their birth. Their offspring are weaned and slaughtered within this same time sequence. From the time of initial breeding to the time the expanded breeding produces pigs for slaughter is typically about twenty months.

The size of the breeding herd essentially determines the number of sows farrowing and the subsequent size of the pig crop. Pigs-per-litter is affected by death loss and the age composition of the breeding herd, as well as advances in production technologies. However, in the short term, death loss, the age composition of the breeding herd, and existing technology are not readily affected by prevailing economic conditions. Thus, in the short run, hog production is entirely based upon the prevailing breeding herd size and breeding decisions.

The investment decisions of producers, in adjustments in their breeding herds, are a key determinant of hog supply. When producers decide to expand their production capabilities, the return from their investment does not appear for nearly twenty months. At the aggregate level, producers also must hold back gilts that would otherwise be slaughtered. This results in a negative supply response in the short run. Hence, this dual role of gilts in the pork production process has direct implications on the cyclical nature of production

(Jarvis 1974; Rosen 1987). However, when producers disinvest by increasing their culling rate of sows, the supply response is immediate.

Biological Restrictions and Supply Response

As Shonkwiler (1982) has noted, the lag lengths in agricultural supply specifications are typically determined by biological factors and impose restrictions on supply response behavior. This is particularly evident in the lag structures in dynamic econometric models of the livestock sector. The biological relationships in the production process also can provide additional information beyond the sequential specification of the supply structure and the implied lag lengths of explanatory variables in the various supply categories.

As noted previously, Johnson and MacAulay (1982) first incorporated biological restrictions in the supply structure of a quarterly beef model. Historical biological relationships between stock and flow categories were used to obtain restrictions on the parameter estimates within the supply structure. Thus, the biological restrictions imposed by nature were used as a priori information to estimate the stock-to-flow and flow-to-flow relationships in production phases. Similar biological restriction are used in the supply structures of the present models.

Because a time lag of approximately five to six months exists between birth and slaughter, a strong relationship should exist between the pig crop (PCUS) lagged two quarters and barrow and gilt slaughter (BGSUS). A similar relationship should hold between the pig crop lagged two quarters and additions to the breeding herd (ABHUS). Also, a fairly constant proportion of the sows in the breeding herd (BHUS) is sent to slaughter (SSUS) each quarter because of the continual aging process, which results in productivity declines. Given constancies in breeding practices, a relationship should hold for the number of sows in the breeding herd (BHUS) and the number of sows farrowing (FARROW). Finally, sows farrowing (FARROW) is highly correlated with the pig crop (PCUS).

These basic biological relationships can be synthesized in ratios of the underlying stock-to-flow and flow-to-flow categories. Quarterly means and standard deviations of these biological ratios are given in Table 4.1 for the sample period 1970 to 1986. These ratios suggest the average relationships between the stock-to-flow and flow-to-flow categories for each quarter. They also indicate the seasonality in the pork production process. The first ratio suggests that an average of 3 to 8 percent of the total pig crop is added back into the breeding herd. The second ratio implies that an average of 12 to 15 percent of the sows in the breeding herd are slaughtered. Similarly, the third ratio suggests that 32 to 43 percent of the breeding herd farrows each quarter. The fourth ratio indicates that an average of more than seven pigs is produced per sow. Finally, the fifth ratio suggests that an average of 80 to 93 percent of the pig crop is slaughtered each quarter. These average relationships between these supply and inventory categories form the underpinnings of the biological restrictions.

The ratios in Table 4.1 give average relationships among the supply and inventory categories. These relationships are not expected, a priori, to be stationary over the sample period. Plots of the ratios by quarter against time indicate trends and structural shifts in the ratios. For example, the ratio $ABHUS_t/PCUS_{t-1}$ trends downward until 1974 in the first two quarters. The ratio $SSUS_t/BHUS_{t-1}$ trends upward after 1975 during all four quarters. Also, the number of pigs per litter, represented by the ratio $PCUS_t/FARROW_t$, exhibits a similar upward trend after 1975 with increases in sow productivity. Similar trends and threshold points were found in the remaining ratios.

Table 4.1.   Quarterly means of stock-to-flow and flow-to-flow ratios
from the first quarter of 1970 through the third quarter
of 1986

Ratio	Quarter			
	1	2	3	4
$ABHUS_t/PCUS_{t-2}$	0.0434	0.0459	0.0807	0.0364
	(0.0165)[a]	(0.0222)	(0.0191)	(0.0067)
$SSUS_t/BHUS_{t-1}$	0.1212	0.1260	0.1539	0.1484
	(0.0109)	(0.0146)	(0.0228)	(0.0142)
$FARROW_t/BHUS_t$	0.3289	0.4256	0.3626	0.3681
	(0.0248)	(0.0201)	(0.0318)	(0.0308)
$PCUS_t/FARROW_t$	7.1637	7.3971	7.3183	7.3193
	(0.2308)	(0.1974)	(0.1953)	(0.2299)
$BGSUS_t/PCUS_{t-2}$	0.8764	0.8877	0.9319	0.7984
	(0.0304)	(0.0282)	(0.0409)	(0.0633)

[a]Standard deviations are in parentheses.

The trends can occur because of a number of factors, including changes
in market conditions, technological advancements, and structural shifts. Some
influences include increases in sow productivity, improved management and
feeding practices, and changes in seasonality due to the adoption of
confinement units. The trends in these ratios can be exploited in a simple
time-varying parameter context. The trends in the ratios can be incorporated
as prior restrictions by

$$R_i = a_i + b_i * T65 * DV + c_i * DV + e_i , \quad i = 1, 2, 3, 4 \quad (4.1)$$

where R_i is the quarterly ratio, T65 is a time trend, DV is the threshold point,
e_i is an error term with the usual assumptions, and a_i, b_i, and c_i are the
parameters to be estimated. These parameter estimates assimilate the implied
technological and structural changes that have occurred among the biological

stock-to-stock and flow-to-flow categories. But imposing the parameter estimates from (4.1) in the underlying supply equation implies that the forces that cause these structural breaks and trends are exogenous to the model. The estimation results for the quarterly ratios that incorporate the structural shifts and trends are included in Chapter 5.

Rational Expectation Model Structure

The quarterly rational expectation pork industry model contains nine behavioral equations and four identities. The quarterly time frame is used to accurately depict the dynamics of the production process and the role of seasonality in determining supply and demand. Also, the quarterly time frame is used for the *Hogs and Pigs* report.

The quarterly pork model provides behavioral representations of the major components of the industry supply and demand structure. The supply structure provides a disaggregated characterization of the phases in the production process. The supply block includes behavioral relationships for additions to the breeding herd, sow slaughter, sows farrowing, the pig crop, and barrow and gilt slaughter. The breeding herd inventory is derived through an identity. The breeding herd inventory, the pig crop, and farrowing are reported in the *Hogs and Pigs* report. These relationships in the supply structure are illustrated in Figure 4.2.

Domestic pork production is the sum of the two slaughter categories, multiplied by their respective slaughter weights. The slaughter weights for barrows and gilts and sows respond to changes in input and output prices.

Domestic pork production, which is in live weights, is transformed into commercial pork production by using a behavioral equation. Commercial production is in carcass-weight equivalent. Per capita consumption, in retail weights, is obtained through an identity. The price determination of the

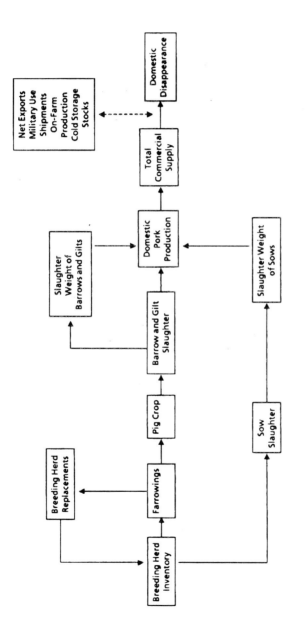

Figure 4.2. Supply components of the rational expectation model

model is contained in the retail demand structure. Price is determined at the retail level. The behavior of meat processors is captured in a retail-farm-margin behavioral equation. The farm price, derived from identity, is defined as between the retail price and the margin. Cold storage stocks, exports and imports, on-farm production, shipments, and military use are considered exogenous.

Supply Structure

As with other livestock models, the supply component is specified to reflect the hog production process. The biological nature of hog production provides constraints on producers' response to economic variables. The biological timetable of production is used to define appropriate lag lengths in the supply relationships. Also, the historical relationships among appropriate supply and inventory categories are used to provide additional restrictions on supply movements. The equations for additions to the breeding herd, sow slaughter, sows farrowing, pig crop, and barrow and gilt slaughter are restricted by these biological restrictions. The supply structure closely resembles previous work of Blanton (1983), Oleson (1987), and Skold, Grundmeier, and Johnson (1988).

The biological time frame of the pork production process also is used to define the planning horizon of producers. This, in turn, defines the length of the forward-looking expectations. The profitability expectations of producers enter into their breeding herd investment decisions; specifically, in the equations that represent additions to the breeding herd and sow slaughter. The profitability expectations are represented by a parsimonious set of conditioning variables: the expected prices of barrows and gilts and feed and the real interest rate.

As depicted in Figure 4.2, the supply of hogs going to market begins with determining the size of the breeding herd. The investment decision of producers is represented by an equation representing additions to the breeding herd (ABHUS), specified as

$$ABHUS_t = f(PCUS_{t-2},\ {}_tFPPK_{t+2},\ {}_tFC_{t+2},\ RIFCL_t)\ , \qquad (4.2)$$

where $PCUS_{t-2}$ is the pig crop lagged two quarters, ${}_tFPPK_{t+2}$ is the expected farm price of barrows and gilts in period $t + 2$, ${}_tFC_{t+2}$ is the expected cost of feed in period $t + 2$, and RIFCL is the real interest rate. The two-quarter lag in the pig crop approximates the age at which gilts enter the breeding herd. The parameters on the lagged pig crop are restricted to incorporate biological restrictions.

The farm price and feed cost represent the profitability expectation of producers. Feed costs are composed of corn and soymeal prices, weighted to reflect a typical ration. Feed costs are the major variable cost in farrow-to-finish operations (Van Arsdall and Nelson 1984). The expectations horizon on the farm price and feed costs reflects the length of time to feed pigs to slaughter weights. Thus, producers retain gilts if the expected profits resulting from the sale of their offspring will provide a satisfactory return. The real interest rate reflects the investment cost of retaining gilts.

Disinvestment in the breeding herd inventory is represented by an equation for sow slaughter (SSUS), specified as

$$SSUS_t = f(BHUS_{t-1},\ {}_tFPPK_{t+2},\ {}_tFC_{t+2},\ RIFCL_t)\ , \qquad (4.3)$$

where $BHUS_{t-1}$ is the breeding herd lagged one quarter. The breeding herd lagged one quarter represents the stock of sows available for slaughter. The sows are generally slaughtered after farrowing and weaning. Again, the parameters on the lagged breeding herd are restricted by prior information.

The same set of conditioning variables with the same expectation horizon as in the additions to the breeding herd equation (4.2) are contained in the sow slaughter equation. Similarly, producers are assumed to keep sows in the breeding herd if the profitability expectations of their subsequent output yields a sufficient return. The interest rate is included to represent the opportunity cost of investing in replacement gilts. A significant portion of sow slaughter is simply determined by culling due to the aging of the breeding herd. Thus, culling due to declines in sow performance is determined primarily by biology, not economic factors. Consequently, no contemporaneous output and input prices are included in the specification.

The additions to the breeding herd and the level of sow slaughter are inflows and outflows, respectively, into the breeding herd inventory. The identity that determines the breeding herd inventory (BHUS) represents this stock flow relationship. The relationship between stocks and flows is based on the identity (Blanton 1983)

$$CI_t + S_t = CI_{t-1} + IN_t , \qquad (4.4)$$

where CI_t is the closing inventory, S_t is the outflow or slaughter, and IN_t is the inflow from one stage to another. The beginning inventory is CI_{t-1}. By rearranging the identity, it is clear that the change in the inventory is equal to the difference between inflows and outflows. This same identity applies to the breeding herd inventory relation. The breeding herd stock is determined by the carry-in inventory and the inflows (additions) and the outflows (slaughter):

$$BHUS_t + SSUS_t = BHUS_{t-1} + ABHUS_t . \qquad (4.5)$$

With simple manipulation, this stock flow relationship obtains the identity that determines the breeding herd inventory.

The breeding herd inventory represents the stock of sows available for breeding. The sows that are bred and subsequently farrow determines sows farrowing (FARROW),

$$FARROW_t = F \ (BHUS_t \ ) \ . \tag{4.6}$$

Sows farrowing is assumed to be strictly dependent on the size of the breeding herd, and is constrained by biological restrictions. The timetable for sows to reenter the breeding herd is thus assumed to be dependent only on the regularities in the production process, not on economic variables.

Farrowing, in turn, determines the size of the pig crop (PCUS). Pigs per litter is affected by death loss and the underlying age distribution of the breeding herd, as well as by advances in technology and management practices. However, death loss, the age composition of the breeding herd, and existing technology are not readily affected by prevailing economic conditions in the short term. Thus, hog production is assumed to be entirely based upon the number of sows that farrowed. Sows farrowing, in turn, depends on the prevailing breeding herd size and breeding decisions. Omitting technological advances and regularities in the seasonality of farrowing, this implies a constant relationship between the prevailing size of the breeding herd and the pig crop. The pig crop is specified accordingly,

$$PCUS_t = f \ (FARROW_t \ ) \ . \tag{4.7}$$

The parameters in (4.7) are restricted by biological accounting ratios.

Barrow and gilt slaughter (BGSUS) is limited by the number of pigs grown to slaughter weights. Producers can sell market hogs at heavier or lighter weights, but barrow and gilt slaughter is limited by the previous pig crop. Given regularities in feeding practices, this technical relationship is specified as

$$BGSUS_t = f\ (PCUS_{t-2})\ .$$ (4.8)

The two-quarter lag in the pig crop represents the five to six months required to finish a pig to slaughter weights.

The fundamental determinant of the hog supply is the size of the breeding herd. In the short run, however, producers have discretion in adjusting marketing times. These short-run supply adjustments are represented in equations that determine the live weight of barrows and gilts (LWBG) and the live weight of sows (LWS). These market weight equations are specified as

$$LWBG_t = f\ (FPPK_t\ ,\ FC_t\ ,\ T65_t\ ,\ D_i\ )$$ (4.9)

and

$$LWS_t = f\ (FPPK_t\ ,\ FC_t\ ,\ T65_t\ ,\ D_i\ )\ .$$ (4.10)

Producers are assumed to respond to current market conditions in determining the time of sale. Thus, only contemporaneous output and input prices are included in the slaughter weight specifications. The time trend ($T65_t$) is included to reflect meat packers' demand for heavier carcasses. This, in part, is a result of improved carcass composition due to breeding advancements. The quarterly dummy variables (D_i) account for seasonality in market weights due to influences of weather and other factors.

Domestic pork production (PPF) is derived through an identity that equals the sum of barrow and gilt slaughter and sow slaughter, multiplied by their respective average slaughter weights

$$PPF_t = BGSUS_t\ *\ LWBG_t\ +\ SSUS_t\ *\ LWS_t\ .$$ (4.11)

Boar slaughter is not explicitly introduced in the identity because it is a minor component of total slaughter. Domestic pork production, in

liveweights, is transformed into carcass weights in the equation that
determines commercial pork production (TOTSPK). Commercial pork
production is specified as

$$TOTSPK_t = f\,(PPF_t\,,\, LT65_t\,)\,, \tag{4.12}$$

where $LT65_t$ is the logarithm of the time trend. This logarithm captures the
carcass weight improvements, more usable carcass per pound of liveweight
hogs marketed. The coefficient on domestic pork production incorporates
boar slaughter, which is typically proportional to domestic pork production.

Demand and Price Determination Structure

The price determination is assumed to occur at the retail level. Fox
(1953) observed that livestock production is essentially fixed in the short run,
and, hence, the determination of retail price depends on the location of the
demand curve. The retail price is linked to the farm-level price through a
margin equation. This structure simplifies the retail farm linkage by
circumventing the wholesale market.

A relatively simple retail demand specification is used. It does not
conform to the tenets of consumer behavior, namely the Slutsky condition and
the ability to integrate, but does include the fundamental determinants of the
demand for pork. Because of the computational burden inherent in nonlinear
rational expectation models, the almost-assured rejection of the Slutsky
conditions by the data (see Johnson et al. 1986), and the partial equilibrium
nature of the pork market model, this simple demand specification was
chosen. The retail demand is specified in price-dependent form. The retail
price of pork (RPPK) is specified as

$$RPPK_t = f\ (PCPK_t\ ,\ RPBF_t\ ,\ FEXP_t\ ,\ T65_t\ ,\ D_i\ )\ ,\qquad (4.13)$$

where $PCPK_t$ is per capita consumption of pork, $RPBF_t$ is the retail price of
beef, $FEXP_t$ is per capita food expenditures, $T65_t$ is a time trend, and D_i
values are the quarterly dummy variables. The retail prices of pork and beef
and food expenditures are deflated by the consumer price index. Only beef is
included as a substitute good, which assumes separability between pork and
beef consumption and other commodities. Other meat commodities, such as
chicken, are not included as substitutes because of the usual insignificant
relationship with pork demand (Moschini and Meilke 1988). Food
expenditures capture the effect of changes in income in the assumed two-stage
budgeting process completed by consumers. The time trend is a proxy for
other secular and temporal factors that affect the demand for pork. The
quarterly dummy variables capture the seasonality in pork retail prices.

The retail price of pork is linked to the farm price through a margin
equation. The specification of the retail farm margin (MARGIN) follows
previous work by Wohlgenant and Mullen (1987). Wohlgenant and Mullen,
following Gardner (1975), note that changes in the margin can originate from
supply and demand movements and from changes in marketing costs. Shifts
in demand and supply influence the retail-farm margin through the quantity of
output processed by packers and the changes in the retail price. The retail-
farm margin is specified as

$$MARGIN_t = f\ [RPPK_t\ *\ D_i\ ,\ RPPK_t\ *\ (TOTSPK_t\ /\ POP_t\ )\ ,\ MKTCST_t\ ]\ ,$$

$$(4.14)$$

where $RPPK_t$ is the retail price of pork, $TOTSPK_t$ is commercial pork
production, POP_t is the U.S. population, and $MKTCST_t$ is an index of packer
marketing costs. $MKTCST_t$ is the simple average of the index of earnings of

employees in packing plants and the producer price index of fuels and related power products. Again, D_i values are quarterly dummy variables. The retail-farm margin, the retail price of pork, and the index of packer marketing costs all are deflated by the consumer price index.

In this formulation, changes in the retail price do not have a constant relationship with the retail-farm margin. The effects of movements in supply enter through $RPPK_t * (TOTSPK_t/POP_t)$, the per capita retail value of commercial pork production. The farm price of barrows and gilts (FPPK) is simply the retail price less the retail-farm margin. It is derived by the identity

$$FPPK_t = (RPPK_t - MARGIN_t) * CPI_t , \qquad (4.15)$$

where CPI_t is the consumer price index.

The market clearing identity equates pork supply and demand. From this identity, domestic disappearance (TOTDPK) is obtained,

$$TOTDPK_t = TOTSPK_t + OTHER_t , \qquad (4.16)$$

where $TOTSPK_t$ is commercial pork production and $OTHER_t$ incorporates the other uses and supply flows. Included in the latter are exports and imports, net cold storage stock changes, shipments, military use, and on-farm production. All of these categories are considered exogenous. Domestic disappearance divided by the U.S. population and multiplied by the carcass-retail conversion ratio yields per capita pork consumption (PCPK),

$$PCPK_t = (TOTDPK_t / POP_t) * PVERT. \qquad (4.17)$$

The carcass-retail conversion ratio (PVERT) has increased with improved breeding, slaughter, and packing practices.

Futures Market Expectation Model Structure

Futures market prices are the other expectation mechanism used in the current study. Using futures market prices as proxies for the expectations of individuals removes the price determination from the structure of the model. Thus, only the specification of the supply structure needs to be modified, and the demand structure can be disregarded. The periodicity of the data remains quarterly.

Futures market prices are posited in the additions to the breeding herd (4.2) and sow slaughter (4.3) equations in the supply structure. Additions to the breeding herd and sow slaughter are the inflows and outflows, respectively, and determine changes in the breeding herd inventory (4.5). The sows farrowing (4.6) and pig crop (4.7) equations remain specified as technical relationships. The biological restrictions remain imposed in these supply equations. The rest of the supply and demand block is discarded because price determination is removed from the model's structure. Only the supply equations that predict the selected categories from the *Hogs and Pigs* report are needed.

Expectations are unobservable, and, consequently, the timing of the formation of expectations is unknown. This allows some discretion in the selection of the futures contracts that represent the planning horizon of producers. However, some information is obtained from the timetable of hog production. The period between breeding and subsequent slaughter of the offspring is approximately ten months. Thus, price expectation for the barrows and gilts price should be the futures contract for live hogs at slaughter ten months from the time of the breeding decision.

This general expectation horizon was used by Miller and Kenyon (1980) in a sows farrowing equation. Miller and Kenyon also included lagged cash prices for barrows and gilts and corn and soymeal in their specification.

Including both cash and futures market prices induces possible multicollinearity problems. Also, they ignored the availability of futures market prices for the corn and soymeal. Futures market quotes for these input prices also could serve as proxies for profitability expectations.

In this study, to ease the data collection burden, only the futures market prices of corn are used as the expectation for anticipated feed costs. The majority of feed costs for farrow-to-finish operations is the cost of corn (USDA 1986). A forecast horizon similar to the expected price of barrows and gilts is used for the corn price. However, the horizon is shifted back because feeding occurs before slaughter. Thus, the expectations are assumed to be formed at approximately the time of breeding for the period the subsequent offspring would be fed to market weights.

Live hog futures contracts are traded at the Chicago Mercantile Exchange (CME) for February, April, June, July, August, October, and December. The contracts used as proxies for expectations for barrow and gilt prices are assumed to be formed two quarters before the birth of the offspring, reflecting the breeding and gestation period. The live hog contract used is the contract approximately ten months from this date, which represents the approximate time of slaughter. In Table 4.2, the live hog contracts used for the four quarters are presented. The expected price of barrows and gilts is a simple quarterly average of the closing prices of the contract at the approximate time of the breeding decision.

Corn futures contracts are traded at the Chicago Board of Trade (CBT) for March, May, July, September, and December. Similarly, closing price quotes from these corn futures contracts represent the expected cost of feed during the finishing period. Again, the expectations are formed two quarters before farrowing and are simple quarterly averages of the corn contracts. In Table 4.2 the corn contracts designated as the expectation proxies for each

Table 4.2. Designation of live hog and corn futures contracts used as the price expectation proxies by quarter

Quarter	Live Hog Contract Month	Corn Contract Month
1	June	March
2	October	July
3	December	December
4	April	March

quarter are given. These contracts represent the expectations of the corn price during the approximate time of feeding.

The futures contracts for live hogs and corn serve as proxies for the profitability expectations of producers in the equations for additions to the breeding herd (ABHUS) and sow slaughter (SSUS). The specification for these two equations with futures market based expectations are

$$ABHUS_t = f\,(PCUS_{t-2}\,,\,FUTHOG_t\,,\,FUTCORN_t\,,\,RIFCL_t\,)\,, \quad (4.18)$$

and

$$SSUS_t = F\,(BHUS_{t-1}\,,\,FUTHOG_t\,,\,FUTCORN_t\,,\,RIFCL_t\,)\,, \quad (4.19)$$

where $FUTHOG_t$ is the quarterly average of the closing live hog futures contract and $FUTCORN_t$ is the quarterly average of the closing corn contract (see Table 4.2 for details). The biological restrictions are retained. Thus, the relationships between $ABHUS_t$ and the pig crop lagged two quarters ($PCUS_{t-2}$), and $SSUS_t$ and the lagged breeding herd ($BHUS_{t-1}$) are constrained by prior information from historical and biological patterns. The real interest rate (RIFCL) is also kept in the specifications as a proxy for the cost of credit.

Summary

The econometric models of the pork sector provide a method to integrate alternative sets of information in developing the estimates for the *Hogs and Pigs* report. The models incorporate known biological relationships as prior information. The biological constraints provide a means to unify short-term supply behavior with the long-run formation of supply response. This feature forces a consistency between past supply and inventory levels and the predictions of the estimates of the breeding herd inventory, sows farrowing, and the pig crop. This consistency is of importance in developing the initial estimates in the *Hogs and Pigs* report.

These econometric models also integrate alternative mechanisms to generate expectations. The first form of expectations used follows the rational expectation hypothesis. Rational expectations are based on the structure and behavioral characteristics of the model, which represents the pork subsector. Thus, the information set on which expectations are based is the biology-governed behavior of supply response, combined with the response of producers, packers, and consumers to changing market conditions.

The second form of the econometric model employs expectations derived from futures market price quotations. Thus, the information set on which production decisions are made is aligned with the anticipations of futures market participants. Futures market expectations are generated outside the model structure and enhance the information set beyond that of rational expectations and survey data.

Combining the estimates through composite forecasting of the breeding herd inventory, sows farrowing, and the pig crop from the rational expectation and futures market expectation models and from the USDA survey provides an efficient, consistent, and precise way to expand the information base of the *Hogs and Pigs* report.

Chapter 5

Estimation Methods
and Results

The econometric models of the pork sector incorporate rational and futures market expectations. The model structures integrate biological restrictions that govern the phases of pork production. Thus, the pork industry models use information from known biological relationships in the production process and incorporate alternative expectation mechanisms that characterize the information processing abilities of individuals.

This chapter presents estimation results and validation statistics for the rational and futures market expectation models of the pork sector. The solution and estimation technique used for the rational expectation, called the Extended Path (EP) method, was developed by Fair and Taylor (1983) and is applicable to rational expectation models with dynamic and nonlinear variables. Imposing biological restriction in the supply block introduces nonlinearities in the variables of the rational expectation model of the pork sector. These nonlinearities require alternative estimation procedures, different from those described in Chapter 3 for linear rational expectation models.

Estimation of Nonlinear Rational Expectation Models

The structures of econometric models of the livestock sector in general, and pork models specifically, have been nonlinear. Cycles in prices and production necessitate the adoption of nonlinear structures (Okyere and Johnson 1987). Adequately capturing the short- and long-run livestock supply response also often requires a nonlinear structure. Furthermore, incorporating prior information, such as biological restrictions, forces nonlinearities in the variables.

Until recently, computation methods were only available for linear rational expectation models, as described in Chapter 3. Consequently, applications and scope of econometric models that integrate the rational expectation hypothesis (REH) have been limited. This is particularly true for agricultural sector applications. Fair and Taylor (1983) have proposed a general solution and an estimation method for nonlinear rational expectation. The computation burden and cost of their EP method has limited its use, but it has been successfully applied by Fair and Taylor (1983), Fair (1984), and Holt and Johnson (1988), and unsuccessfully by Moore (1985).

Overview of Solution and Estimation Method

Consider a general dynamic and nonlinear rational expectation model,

$$f_i\ (y_t,\ ...,\ y_{t-p},\ E_t y_t,\ ...\ E_t\ y_{t+h},\ X_t, \alpha_i) = u_{it}\quad i = 1,\ ...,\ n, \tag{5.1}$$

where y_t is an n-dimensional vector of the endogenous variables at time t, x_t is a vector of the exogenous variable at time t, E_t is the conditional expectation operator based on the model and on information through period t, α_i is a vector of parameters, and u_{it} is a stationary random disturbance with mean zero. The random disturbance may be correlated across equations ($Eu_{it}u_{jt} \neq 0$ for $i \neq j$) and over time ($Eu_{it}u_{is} \neq 0$ for $t \neq s$).

If the model were linear, the rational expectation would be obtained by solving for the reduced form of the system of equations and eliminating the expectation variables through substituting expectations of the exogenous variables. The rational expectation could then be written explicitly as a function of observable variables.

For nonlinear models, the reduced form typically cannot be calculated analytically, but rather can be evaluated numerically. The EP method developed by Fair and Taylor (1983) numerically solves for the rational expectation through a series of Gauss-Seidel iterations. In brief, for a given parameter vector, initial guesses are made of the values of $E_t y_{t+j}$ for $j = 1$, ..., J. A series of Gauss-Seidel iterations is then completed to obtain a new path for $E_t y_{t+j}$. These solution values replace the initial guesses, and another set of Gauss-Seidel iterations is completed until convergence is achieved. The path is then extended from J to $J + 1$, and the process is repeated until some convergence level is attained. If values of $E_t y_{t+j}$ for the paths J and $J + 1$ are within some tolerance level, rational expectation is solved for the initial parameter vector. If the tolerance criterion is not met, the procedure is repeated for $J + 2$ and so forth. Thus, the path of rational expectation begins with an arbitrary path and is successively extended until it converges with the previous path.

The solution method yields the rational expectation, consistent with the model's structure, given initial estimates of the parameter vector. To estimate the parameter vector, Fair and Taylor (1983) propose replacing the calculated reduced form with a numerical approximation from the nonlinear model. This numerical solution replaces the analytic and restricted solution from the linear case. The reduced form can then be evaluated with full information estimation techniques. Full information estimators are appropriate because of the cross-equation restriction implied by the REH. Accordingly, under the assumption that the disturbance is distributed normally, the reduced form of the structural parameters can be evaluated with numerical maximum likelihood routines.

Solution Method

The method solves the nonlinear rational expectation model (5.1) for a given set of parameters α_i. The model is assumed to have no serial correlation in the disturbance terms. The solution method can be modified if serial correlation is present in the disturbance terms, but this adds to the computation difficulty and cost. Also, conditional expectations in (5.1) are approximated by setting the future disturbances to their conditional means. This results in an essentially deterministic simulation to solve for the rational expectation. Fair and Taylor (1983) present an alternative stochastic simulation method that yields more accurate solutions. Again, this latter method significantly increases computation costs.

The exogenous variables, x_t, and the expected values of the exogenous variables, $E_t x_{t+j}$ for period $t + j$ (based on information from period t) are assumed to be known for all t and j. The expected values of the exogenous variables that are generated from a stochastic process can be treated as exogenous in the estimation procedure. For this study, the expected values of the stochastic exogenous variables are generated from univariate, autoregressive-integrated, moving-average (ARIMA) processes. These processes are considered completely determined outside the structure of the model.

Deriving the rational expectation begins with an arbitrary guess at the expected endogenous variables. Following the notation of Fair and Taylor (1983), let the initial guess of the expected endogenous variables, $E_t y_{t+r}$, be represented as g_r, for $r = 0, 1....$ Only a finite set of these will be used to converge to a finite tolerance level. Let h be the planning horizon of individuals in the systems, and let k be an integer representing the number of periods beyond h that the model must solve to reach a tolerance criterion δ.

The solution method is described by Fair (1984, 372-73) in five steps:

1. Set $E_t y_{t=r}$ equal to g_r, $r = 0, 1, ..., k + 2h$. Call these solution values $e_r(1, k)$, $r = 0, 1, ..., k + 2h$, and the solution values for subsequent iterations $e_r(i, k)$, $i > 1$.

2. Obtain a new set of values for $E_t y_{t+r}$, $r = 0, 1, ..., k + h$, by solving the model dynamically for y_{t+r}, $r = 0, 1, ..., k + h$. This is done by replacing $E_t x_t, ..., E_t x_{t+h+k}$ with the forecasts or actual values of $x_t, ..., x_{t+h+k}$, and using values $e_r(1,k)$ in place of $E_t y_{t+r}$. Call these new guesses $e_r (i + 1, k)$, $r = 0, 1, ..., k + h$. The solution for each period requires a series of Gauss-Seidel iterations until a convergence criterion σ is obtained. This series of iterations is called Type I.

3. Compute the absolute difference between the new guess $e_r (i + 1, k)$ and the previous guess $e_r(i, k)$ for each expectation variable and for each period, $r = 0, 1, ..., h + k$. If any of the absolute differences $e_r (i + 1, k)$ and $e_r(i, k)$ is more than a tolerance criterion ϵ, then return to step 2, and increase i by 1. If convergence is achieved, move to step 4. Call the solution to this series of Gauss-Seidel iterations $e_r(k)$, $r = 0, 1, ..., k + h$. This series of Gauss-Seidel iterations in steps 2 and 3 is called Type II.

4. Repeat steps 1 through 3 by advancing k to $k + 1$. Compute the absolute difference between $e_r(k + 1)$ and $e_r(k)$ for each element for $r = 0, 1, ..., h$. If the difference is no greater than δ, then proceed to step 5, otherwise increase k by 1 and repeat steps 1 through 4. Call this iteration Type III. Let e_r be the convergent vector from the Type III iteration.

5. Use e_r for $E_t y_{t+r}$, $r = 0, 1, ..., h$, and the actual values for x_t to solve the model for period t.

The convergence criterion for Type II iterations should be less than the overall tolerance level (i.e., $\epsilon < \delta$). Similarly, the Type I tolerance level σ is less than the Type II tolerance level ϵ. As Fair and Taylor (1983) note, the computational costs depend on the number of passes through the model to achieve convergence. A pass is defined as a single evaluation of the left variables as a function of the right variables. Let N_1 be the number of Type I iterations, N_2 be the number of Type II iterations, and N_3 be the number of Type III iterations required for convergence. The total number of passes through the model for Type III convergence is given by

$$\sum_{q=k}^{k+N_3-1} [N_2 * N_1 * (h+q+1)] . \qquad (5.2)$$

The computational cost depends on the planning horizon h and, of course, the general structure of the model. As with most numerical procedures, there is no guarantee that any iteration will converge. Nonetheless, if convergence is reached along the series of extended paths, the rational expectation consistent with the structure of the model is obtained.

Full Information Estimation Procedure

Full information estimators are required to fully exploit and incorporate the cross-equation restrictions implied by the REH. Fair and Taylor (1983) outline the full information maximum likelihood (FIML) procedures for the general model structure (5.1). Assume that the first m equations in (5.1) are stochastic and the remaining n - m equations are nonstochastic ($u_{it} = 0$, m + 1, ..., n).

Let J_t be an n * n Jacobian matrix whose ij element if $\partial fi/\partial yjt$ for i, j + 1, ..., n, and let S be an m * m matrix whose ij element is $(1/T) \sum_{t=1}^{T} u_{it} u_{jt}$ for i, J = 1, ..., m. The unknown parameters are α. If the u_{it} are normally and independently distributed, then the FIML estimates of α are obtained by maximizing the log-likelihood function, with respect to α. The estimate of the covariance matrix of these estimates is given by

$$L = - \frac{T}{2} \log \mid S \mid + \sum_{t=1}^{T} \log \mid J_t \mid, \tag{5.3}$$

$$\dot{V} = - (\frac{\partial^2 L}{\partial \alpha \, \partial \alpha'})^{-1} , \tag{5.4}$$

where the derivatives are evaluated at the optimum.

Thus, the solution from the EP for a given α for the nonlinear rational expectation model (5.1) gives the values for $E_t y_t$, ..., $E_t y_{t+h}$. By using data on y and x, the values of u_{it} can be computed for t = 1, ..., T. Then the matrix S and the Jacobian matrix can be computed, which completes determination of the log-likelihood function. The FIML estimates then can be obtained by maximizing L with respect to α by using numerical procedures.

Numerical Optimization Procedure

Nonlinear optimization problems, such as the maximization of the log-likelihood function (5.3) with respect to α, require iterative, numerical techniques. In general, numerical optimization techniques begin with initial parameter estimates and then repeatedly compute new estimates until some convergence criterion is reached. Many numerical optimization algorithms are available. They basically differ by the rules that govern the parameter search, resulting in an improved value of the objective function. In this study, the quasi-Newton method called Davidson-Fletcher-Powell (DFP) was used to obtain the FIML parameter estimates of the rational expectation model.

Following Judge et al. (1980, 729-35), the DFP algorithm can be explained by considering the objective function $H(\theta)$. For example, $H(\theta)$ may be the negative of the log-likelihood function (5.3). Assume all parameters, including the variance-covariance parameters, are contained in a Kx1 vector θ.

The objective is to find a sequence θ_1, θ_2, ..., θ_n of vectors in parameter space such that θ_n minimizes H(θ) approximately.

For a given point θ_n in parameter space, a step direction δ is chosen so that the objective function H(θ) declines. The distance moved is controlled by the step length t. The step direction δ and step length t are chosen such that

$$H(\theta_n + t\delta) < H(\theta_n). \tag{5.5}$$

To minimize H(θ), δ should be such that H(θ_n + tδ) is a decreasing function of t for t close to zero. Thus, for a given δ,

$$\frac{d\,[H\,(\theta n + t\delta)\,]}{dt}\,\Big|\,t{=}0 = \left[\frac{\partial H}{\partial \theta}\,\Big|\,\theta_n\right]'\left[\frac{d\,(\theta n + t\delta)}{dt}\,\Big|\,t{=}0\right] = \left[\frac{\partial 4}{\partial \theta}\,\Big|\,\theta n\right]'\delta \tag{5.6}$$

must be less than zero. Let the gradient of the objective function

$$\frac{\partial H}{\partial \theta}\,\Big|\,\theta_n \tag{5.7}$$

be denoted as γ_n. The step direction is chosen by

$$\delta = -P_n\,\Upsilon_n, \tag{5.8}$$

where P_n is any positive definitive matrix. Then for the nth iteration

$$\theta n + 1 = \theta n - t_n P_n \Upsilon_n, \tag{5.9}$$

where t_n is the step length for the nth iteration. The positive definite matrix P_n defines the step direction. For DFP, P_n is defined as

$$P_{n-1} + \frac{\lambda_{n-1}\,\lambda'_{n-1}}{\lambda'_{n-1}\,(\Upsilon_n - \Upsilon_{n-1})} - \frac{P_{n-1}\,(\Upsilon_n - \Upsilon_{n-1})\,(\Upsilon_n - \Upsilon_{n-1})'\,P_{n-1}}{(\Upsilon_n - \Upsilon_{n-1})'\,P_{n-1}\,(\Upsilon_n - \Upsilon_{n-1})} \quad (5.10)$$

for the nth iteration. In (5.10) λ is defined as the step such that $\theta_{n+1} = \theta_n + \lambda$.

The DFP algorithm is called a quasi-Newton method because the Hessian matrix is only approximated. The Hessian matrix is defined as

$$\left[\frac{\partial^2\, H(\theta)}{\partial\theta\;\partial\theta'} \;\Big|\; \theta_n \right]^{-1} . \quad (5.11)$$

With Newton-Raphson methods, the Hessian matrix is explicitly estimated and defines the step direction P_n. With DFP, however, the $n+1$ iteration direction matrix is

$$P_{n+1} = P_n + M_n , \quad (5.12)$$

where M_n is defined as (5.10).

Data Sources

Data for breeding herd inventory, farrowing, and the pig crop were obtained from the publications that contain final estimates of the *Hogs and Pigs* report (USDA 1977a; 1980; 1984). After 1982 the estimates of these categories are not finalized, so these data are from the most recent *Hogs and Pigs* report (USDA 1970-88).

The other primary source of data is *Livestock and Poultry Situation and Outlook* (USDA 1970-86c). Data on retail and farm prices, barrow and gilt slaughter, sow slaughter, commercial production, domestic disappearance, and other supplies and uses were from this source. Data on corn and soybean meal prices are from *Agricultural Prices* (USDA 1970-86a) and the *Feed Situation and Outlook* (USDA 1970-86b), respectively. The real interest rate data were derived from interest rates on feeder cattle loans obtained from the *Agricultural Finance Databook* (Board of Governors of the Federal Reserve System 1982) and the *Agricultural Letter* (Federal Reserve Bank of Chicago 1983-86). The remaining data for the U.S. population, the consumer price index, and the producer price index of fuels and related power were acquired from the *Survey of Current Business* (U.S. Department of Commerce 1970-86b). The index of meat packing plant workers' earning was created from data in *Employment and Earnings* (U.S. Department of Commerce 1970-86a). The producer price index of fuels and related power and the index of meat packing plant workers' earnings were used to create the index of marketing costs. The futures market prices for live hogs and corn were obtained from *The Wall Street Journal* (1970-86). The live hog and corn prices are the closing prices for the futures contracts traded at the Chicago Mercantile Exchange and the Chicago Board of trade, respectively.

The sample constructed contains 68 quarterly observations for the period 1970-86. The *Hogs and Pigs* report gives aggregate U.S. estimates only biannually, in the December and June reports, for only the first and third quarters. Interpolations were made on the quarterly reported fourteen-state (1973 to 1982) and ten-state data (1970-72 and 1983-86) to yield aggregate U.S. estimates for the breeding herd, farrowing, and the pig crop. These interpolations were used in the second and fourth quarters. Further details on the data and interpolation methods are provided in Appendix A.

Biological Restrictions

The historical relationships between stages in the pork production process are used to define the implied lag lengths of explanatory variables and to restrict the parameters in the supply component of the rational and futures market expectations models.

The underlying determinant of changes in the hog supply is adjustments in the breeding herd inventory. The inflows into the breeding herd are represented by additions to the breeding herd (ABHUS), which is constrained by the pig crop (PCUS) of age for possible breeding. The outflows, represented by sow slaughter (SSUS), are closely linked to the breeding herd inventory (BHUS) of the previous quarter. The lagged breeding herd represents the stock of animals available for slaughter. The difference between the inflows (ABHUS) and outflows (SSUS), plus the carry-in breeding herd inventory, determines the current breeding herd inventories. Given constancies in breeding practices, changes in the breeding herd closely follow movements in sows farrowing (FARROW). Sows farrowing, in turn, determines the size of the pig crop. Barrow and gilt slaughter (BGSUS) is essentially determined by the pig crop lagged two quarters, the length of time required to feed a pig to slaughter weight by conventional feeding practices.

These production stages between breeding and subsequent slaughter are synthesized with ratios of these biological relationships. Plotting the ratios by quarter against time indicates several production regimes. The apparent structural shifts in the ratios may be from improved breeding and feeding practices and changes in management practices due to larger confinement units. These and other technologically induced factors are assumed to be exogenous to the modeled structure and, thus, unexplained by explanatory variables such as changes in relative prices, marketing costs, food expenditures, and interest rates.

In the ratio of additions to the breeding herd to the pig crop lagged two quarters ($ABHUS_t/PCUS_{t-2}$), the ratio exhibited a downward trend in the first and second quarters until 1974. An upward trend after 1975 was present in the ratios of sow slaughter to the lagged breeding herd ($SSUS_t/BHUS_{t-1}$) and in the ratio of sows farrowing to the breeding herd inventory ($FARROW_t/BHUS_t$). Beginning in 1973, the ratio of barrow and gilt slaughter to the pig crop lagged two quarters ($BGSUS_t/PCUS_{t-2}$) tended to trend upward in the third and fourth quarters.

Information from these structural shifts and trends in the ratios were incorporated as prior information with a series of simple time-varying parameter regressions. The ratios were regressed against zero-one dummy variables representing structural breaks and time trends. The resulting parameter estimates were restricted in supply components of the rational and futures market expectation models.

When structural breaks and trends did not seem to exist, the biologically restricted parameter was constrained to the sample mean of the quarterly ratio in subsequent estimation. Thus, for example, in the additions to the breeding herd equation, the parameters on pig crop lagged two quarters in the third and fourth quarters were restricted to the sample quarterly average of the ratio $ABHUS_t/PCUS_{t-2}$. The first and second quarter sample means for the ratio $BGSUS_t/PCUS_{t-2}$ are used as prior information in the same manner. The quarterly averages of the biological ratios for the sample period are presented in Table 4.1.

For ratios that seemed to have structural breaks and trends, the estimation results of the regressions of the quarterly ratios against the zero-one dummy variables and time trends are provided in Table 5.1. Ordinary least squares (OLS) was used to estimate the parameters in these time-varying parameter equations.

Table 5.1. Estimation results of the biological ratios by quarter, 1970-86

Quarter	Ratio	Intercept	Dummy Variable[a]	Time Trend[b]	R^2 [c]	D.W.[d]	Equation
3	$ABHUS_t/PCUS_{t-2}$	0.0773 (16.81)[f]	0.1457 (2.43)	-0.0164 (-2.21)	0.34	1.66	(5.13)
4		0.0376 (21.89)	-0.0443 (-1.92)	0.00474 (1.71)	0.26	2.19	(5.14)
1	$SSUS_t/BHUS_{t-1}$	0.1183 (31.12)	-0.03599 (-2.28)	0.00237 (2.68)	0.36	2.59	(5.15)
2		0.1321 (29.02)	-0.0694 (3.62)	0.00348 (3.27)	0.49	2.41	(5.16)
3	$SSUS_t/BHUS_{t-1}$	0.1620 (19.12)	-0.0815 (-2.25)	0.00394 (1.99)	0.28	2.52	(5.17)
4		0.1543 (27.75)	-0.0393 (-1.63)	0.00169 (1.31)	0.20	3.12	(5.18)
1	$FARROW_t/BHUS_t$	0.3041 (56.35)	-0.1455 (-2.46)	0.0653 (3.12)	0.75	2.42	(5.19)
2		0.4272 (57.90)	-0.1977 (2.39)	0.0689 (2.38)	0.29	2.13	(5.20)
3		0.3297 (106.6)	-0.2881 (-8.15)	0.1192 (9.65)	0.95	1.25	(5.21)

Table 5.1. Continued

Quarter Ratio	Intercept	Dummy Variable[a]	Time Trend[b]	R^{2} [c]	D.W.[d]	Equation
1 $PCUS_t/FARROW_t$	7.1332 (111.0)	-3.0897 (-4.38)	1.1142 (4.48)	0.59	2.33	(5.23)
2	7.2540 (157.0)	-2.0741 (-4.01)	0.8109 (4.46)	0.71	1.23	(5.24)
3	7.2009 (154.7)	-2.3626 (-4.44)	0.8941 (4.81)	0.70	1.08	(5.25)
4	7.1814 (172.3)	-3.2102 (-6.61)	1.1969 (7.09)	0.83	2.05	(5.26)
3 $BGSUS_t/PCUS_{t-2}$	0.9529 (44.07)	-0.2324 (-2.18)	0.0756 (1.99)	0.26	2.17	(5.27)
4	0.7202 (57.18)	-0.4374 (-6.89)	0.1932 (8.61)	0.89	2.04	(5.28)

[a]In (5.13) and (5.14), the zero-one dummy variable equals one if the year is earlier than 1974 and zero otherwise. In equations (5.15) through (5.26), the zero-one dummy variable equals one if the year is 1976 or later and zero otherwise. In (5.27) and (5.28) the zero-one dummy variable equals one if the year is 1973 or later and zero otherwise.

[b]The time trend is T65 in equations (5.13) through (5.18). In equations (5.19) through (5.28) the logarithm of T65 is used.

[c]R^2 is the squared correlation coefficient.

[d]D.W. is the Durbin-Watson statistic.

[e]The values in parentheses under the estimated coefficient are the t statistics.

Note: The general form of the biological regressions is $R_i = a_i + b_i * DV_i + C_i * T65_i + e_i$, where R_i is the biological ratio for quarter i, DV_i is a zero-one dummy variable representing the structural break, $T65_i$ is a time trend, and e_i is a stochastic disturbance term with the usually assumed properties. The parameters estimate are a_i, b_i, and c_i.

ARIMA Models

The rational expectation is conditional on forecasts of the exogenous variables. In rational expectation models, forecasts of stochastic variables are typically generated through time trend projections, autoregressive processes, vector autoregressive (VAR) models, and univariate ARIMA models. Also, the realized values of the stochastic exogenous variable have been used as forecasts of the stochastic exogenous variables (Fair and Taylor 1983). Forecasts of nonstochastic exogenous variables usually are assumed to be known with certainty.

In practice, the forecasts of the stochastic variables often are generated outside the estimated system of equations (Wallis 1980). The implicit assumption is that participants within the modeled system have no structural knowledge about the processes that govern these exogenous variables. Thus, the market participants use mechanisms to generate forecasts of the stochastic exogenous variables that do not impose a certain structure a priori.

In this study, ARIMA models are used to provide forecasts of the stochastic exogenous variables. ARIMA models provide a parsimonious way to depict processes that may be nonstationary and have irregular, seasonal, and cyclical components. Also, ARIMA models produce optimal forecasts; no other univariate linear model with fixed coefficients produces agricultural forecasts with smaller mean square forecast errors (Pankratz 1983). Multivariate models, such as VAR models, may produce forecasts with smaller forecast mean square errors, but VAR models have not shown appreciably better forecasting ability compared with their univariate counterparts (Brandt and Bessler 1984) when applied to agricultural data.

ARIMA models are constructed by using a three-stage procedure developed by Box and Jenkins (1976) and later repeated in standard time-series texts such as Pankratz (1983) and Granger and Newbold (1986). The first stage is identification. In this stage, the sample autocorrelation and partial autocorrelation functions are examined and compared with their theoretical counterparts for known processes such as autoregressive and moving average. The need to differentiate the observations due to apparent nonstationarity is evaluated. Tentative models are chosen.

The second stage of the procedure is estimation. By assuming that the process is distributed jointly normal, an exact likelihood function can be derived. The parameters can be obtained by using maximum likelihood procedures. Alternatively, standard nonlinear regression procedures can be used. In this study, for example, the Gauss-Newton algorithm was used to minimize the sum of squares. The software used to estimate the ARIMA models and in subsequent forecasting of the stochastic exogenous variables was RATS (Doan and Litterman 1987).

The third stage is diagnostic checking, which helps determine if the tentative model adequately represents the underlying data. Part of diagnostic checking entails examining the autocorrelation of residuals. If the model is correctly formulated, the estimated residuals are, on average, uncorrelated, and the autocorrelation coefficients all should be statistically zero. A joint test of all the residual autocorrelations of this proposition is the Box-Pierce test with test statistic Q. A Q statistically different from zero indicates that the residual autocorrelations as a set are significantly different from zero. Other diagnostic checks include fitting extra coefficients and then comparing models and residual plots and fitting data subsets.

ARIMA models were developed for stochastic exogenous variables. Stochastic exogenous variables are the consumer price index (CPI), food expenditures (FEXP), feed costs (FC), marketing costs (MKTCST), demand minus supply (OTHER), U.S. population (POP), real interest rate (RIFCL), and retail beef price (RPBF). The estimated ARIMA models for these variables are given in Table 5.2. Along with the asymptotic t-ratios and the standard error of estimated equations, the Box-Pierce Q statistics are provided for each equation. These and other diagnostic checks indicate that the models provide an adequate representation of the data.

The ARIMA models were used to forecast the stochastic exogenous variables. Projections of all dummy variables, time trends, and the carcass-retail conversion factor (PVERT) were assumed to be known with certainty. These forecasts were used in the Extended Path (EP) method to solve for the rational expectation.

The Rational Expectation Model

The solution and estimation method developed by Fair and Taylor (1983) is a costly, but effective, method of obtaining FIML parameter estimates for the dynamic, nonlinear rational expectation model of the pork sector. The estimated model contains ten stochastic equations and five identities. The model was estimated with 68 quarterly observations from 1970 to 1986.

The biological restrictions were imposed as prior information, constrained 51 parameters, and left 36 unconstrained. The estimated parameters are provided in Table 5.3 with the entire structure of the pork sector model. The 36 parameters estimated with FIML methods are accompanied with their asymptotic t-ratios, below the coefficients in parentheses. Also, for selected coefficients, the partial elasticities, evaluated at sample means, are given in brackets. The 51 parameters constrained by the biological restrictions are included in the additions to the breeding herd (5.37), sow slaughter (5.38), sows farrowing (5.40), pig crop (5.41), and barrow and gilt slaughter (5.42) equations.

Table 5.2. Estimated ARIMA models for exogenous variables

Consumer Price Index

$$(1-B)^2 \text{ CPI}_t = (1 - 0.364B - 0.397B^2 + 0.702B^3) \, \epsilon_t{}^a$$
$$\phantom{(1-B)^2 \text{ CPI}_t = (1 } (-4.08)^b \,\, (-4.42) (7.50)$$

$$Q(24)^c = 34.81 \qquad\qquad \text{SE}^d = 1.147 \qquad\qquad (5.29)$$

Food Expenditures

$$(1 - 0.805B + 0.343B^4)(1 - B^4) \text{ FEXP}_t = \epsilon_t$$
$$ (10.69) (-2.98)$$

$$Q(24) = 19.94 \qquad\qquad \text{SE} = 0.026 \qquad\qquad (5.30)$$

Feed Costs

$$(1 - 0.0906B) \text{ FC}_t = 4.753 + \epsilon_t$$
$$ (19.94) (7.34)$$

$$Q(24) = 15.07 \qquad\qquad \text{SE} = 0.514 \qquad\qquad (5.31)$$

Marketing Costs

$$(1 + 0.355B)(1 - B) \text{ MKTCST}_t = \epsilon_t$$
$$ (-3.20)$$

$$Q(24) = 5.66 \qquad\qquad \text{SE} = 0.0976 \qquad\qquad (5.32)$$

Demand Minus Supply

$$(1 + 0.609B + 0.652B^2 = 0.524B^3)(1-B)(1-B)^4) \text{ OTHER}_t$$
$$ (-5.71) (-5.95) (-5.48)$$

$$= (1 - 0.816B^4) \, \epsilon_t$$
$$ (-8.49)$$

$$Q(24) = 13.23 \qquad\qquad \text{SE} = 50.49$$

$$(5.33)$$

Table 5.2. *(continued)*

U.S. Population

$$(1 - 0.416B - 0.275B^2) \ (1-B)(1-B^4) \ POP_t = (1 - 0.836B^4) \ \epsilon_t$$
$$\quad (3.48) \quad (2.34) \qquad\qquad\qquad (10.7)$$

$$Q(24) = 14.93 \qquad\qquad SE = 0.059 \qquad\qquad\qquad (5.34)$$

Real Interest Rate

$$(1 + 0.524B^2 - 0.362B^3 - 0.335B^5)(1 - B) \ RIFCL_t = \epsilon_t$$
$$\quad (-5.11) \quad (3.22) \quad (2.88)$$

$$Q(24) = 25.87 \qquad\qquad SE = 2.06$$
$$\qquad\qquad\qquad\qquad\qquad\qquad\qquad\qquad\qquad (5.35)$$

Beef Retail Price

$$(1 - B) \ RPBF_t = \epsilon_t$$

$$Q(24) = 28.20 \qquad\qquad SE = 0.041 \qquad\qquad\qquad (5.36)$$

---

[a]The variable ϵ_t denotes a white-noise error process.

[b]Asymptotic t-ratios are reported in parentheses.

[c]The Box-Pierce Q statistic calculated from the residual autocorrelation with the number in the parentheses reflecting the degrees of freedom. The 0.05 critical value for the x^2 distributed Q statistic is 36.415 for 24 degrees of freedom.

[d]SE is the standard error of the estimate.

Note: The lag operator B is defined such that $B^k x_t = x_{t-k}$.

Table 5.3. Full information, maximum likelihood estimates of the rational expectation model, 1970-86

Additions to the Breeding Herd

$ABHUS_t = 0.0434 * D1 * PCUS_{t-2} + 0.0459 * D2 * PCUS_{t-2}$

$\quad + (0.0773 + 0.1457 * DL74 - 0.0164 * DL74 * T65) * D3 * PCUS_{t-2}$

$\quad + (0.0376 - 0.0043 * DL74 + 0.00474 * DL74 * T65) * D4 * PCUS_{t-2}$

$\quad + 16.048 * {}_tFPPK_{t+2} - 131.469 * {}_tFC_{t+2} - 17.075 * RIFCL_t$
$\quad\quad (17.17)^a \quad\quad\quad\quad\quad (-114.1) \quad\quad\quad\quad (-1.86)$
$\quad\quad\quad [0.62]^b \quad\quad\quad\quad\quad\quad [0.55] \quad\quad\quad\quad\quad [-0.07]$ (5.37)

Sow Slaughter

$SSUS_t = (0.1183 - 0.03599 * DUM76 + 0.00237 * DUM76 * T65) * D1 * BHUS_{t-1}$

$\quad + (0.1312 - 0.0694 * DUM76 + 0.00348 * DUM76 * T65) * D2 * BHUS_{t-1}$

$\quad + (0.1620 - 0.815 * DUM76 + 0.00394 * DUM\ 76 * T65) * D3 * BHUS_{t-1}$

$\quad + (0.1543 - 0.0393 * DUM76 + 0.00169 * DUM76 * T65) * D4 * BHUS_{t-1}$

$\quad - 2.648 * {}_tFPPK_{t+2} + 20.159 * {}_tFC_{t+2} + 3.624 * RIFCL_t$
$\quad (-3.68) \quad\quad\quad\quad\quad (3.59) \quad\quad\quad\quad\quad (1.51)$
$\quad [-0.10] \quad\quad\quad\quad\quad [0.08] \quad\quad\quad\quad\quad [0.02]$ (5.38)

Breeding Herd Inventory

$\quad BHUS_t = BHUS_t + ABHUS_t - SSUS_t$

Sows Farrowing

$FARROW_t = (0.3041 - 0.1455 * DUM76 + 0.0653 * DUM76 * LT65) * D1 * BHUS_t$

$\quad + (0.4272 - 0.1977 * DUM76 + 0.0689 * DUM76 * LT65) * D2 * BHUS_t$

$\quad + (0.3297 - 0.2881 * DUM76 + 0.1192 * DUM76 * LT65) * D3 * BHUS_t$

$\quad + (0.3389 - 0.3107 * DUM76 + 0.1244 * DUM76 * LT65) * D4 * BHUS_t$

$\quad\quad\quad\quad\quad\quad\quad\quad\quad\quad\quad\quad\quad\quad\quad\quad\quad\quad\quad\quad\quad\quad\quad\quad\quad$ (5.40)

Table 5.3. *(continued)*

Pig Crop

$$PCUS_t = (7.1332 - 3.0897 * DUM76 + 1.1142 * DUM76 * LT65) * D1 * FARROW_t$$

$$+ (7.2540 - 2.0751 * DUM76 + 0.8109 * DUM76 * LT65) * D2 * FARROW_t$$

$$+ (7.2009 - 2.3626 * DUM76 + 0.8941 * DUM76 * LT65) * D3 * FARROW_t$$

$$+ (7.1814 - 3.2102 * DUM76 + 1.1969 * DUM76 * LT65) * D4 * FARROW_t$$

$$(5.41)$$

Barrow and Gilt Slaughter

$$BGSUS_t = 0.8764 * D1 * PCUS_{t-2} + 0.8877 * D2 * PCUS_{t-2}$$

$$+ (0.9529 - 0.2324 * DUM73 + 0.0756 * DUM73 * LT65) * D3 * PCUS_{t-2}$$

$$+ (0.7202 - 0.4374 * DUM73 + 0.1932 * DUM73 * LT65) * D4 * PCUS_{t-2}$$

$$(5.42)$$

Live Weight of Barrows and Gilts

$$LWBG_t = 211.487 + 7.362 * D2 - 4.491 * D3 + 2.722 * D4$$
$$(41.67)\quad (4.73)\qquad (-3.34)\qquad\quad (1.19)$$

$$+ 2.764 * (FPPK_t/FC_t) + 0.0129 * T65_t$$
$$(4.53)\qquad\qquad\qquad (0.10)$$
$$[0.10]$$

$$(5.43)$$

Liveweight of Sows

$$LWS_t = 402.752 + 5.787 * D2 - 16.445 * D3 - 3.875 * D4$$
$$(59.25)\quad (1.25)\qquad (-8.42)\qquad (-1.18)$$

$$+ 3.954 * (FPPK_t/FC_t) + 1.242 * T65_t$$
$$(4.63)\qquad\qquad\qquad (6.63)$$
$$[0.08]$$

$$(5.44)$$

Domestic Pork Production

$$PFP_t = BGSUS_t * LWBG_t + SSUS_t * LWS_t \qquad\qquad (5.45)$$

Commercial Pork Production

$$TOTSPK_t = 0.6542 * (PPF_t/1000) + 75.839 * LT65$$
$$(116.3)\qquad\qquad\qquad (6.89)$$
$$[0.95]$$

$$(5.46)$$

Table 5.3. *(continued)*

Domestic Disappearance

$$TOTDPK_t = TOTSPK_t + OTHER_t \tag{5.47}$$

Per Capita Consumption

$$PCPK_t = (TOTDPK_t/POP_t) * PVERT_t \tag{5.48}$$

Retail Pork Price

$$RPPK_t = 1.2116 - 0.0416 * D2 - 0.0423 * D3 + 0.0546 * D4$$
$$(11.23)\quad(-4.36)\qquad\qquad(4.28)\qquad\qquad(4.13)$$

$$+ 0.4961 * RPBF_t + 0.01299 * FEXP_t$$
$$(12.32)\qquad\qquad(1.08)$$
$$[0.68]\qquad\qquad[0.05]$$

$$- 0.0669 * PCPK_t - 0.00634 * T65 - 0.0979 * D794$$
$$(21.08)\qquad\qquad(0.13)\qquad\qquad(1.24)$$
$$[-1.53]\tag{5.49}$$

Retail Farm Margin

$$MARGIN_t = 0.2378 * D1 * RPPK_t + 0.2371 * D2 * RPPL_t$$
$$(6.78)\qquad\qquad\qquad(6.48)$$
$$[0.35]\qquad\qquad\qquad[0.35]$$

$$+ 0.2506 * D3 * RPPK_t + 0.2507 * D4 * RPPK_t$$
$$(7.50)\qquad\qquad\qquad(7.05)$$
$$[0.37]\qquad\qquad\qquad[0.37]$$

$$+ 0.00358 * (TOTSPK_t/POP_t) * RPPK_t$$
$$(2.01)$$
$$[0.08]$$

$$+ 0.01293 * MKTCST_t + 0.5492 * MARGIN_{t-1}$$
$$(0.13)\qquad\qquad\qquad(1.97)$$
$$[0.04]\qquad\qquad\qquad[0.55]\tag{5.50}$$

Farm Price of Barrows and Gilts

$$FPPL_t = (RPPK_t - MARGIN_t) * CPI_t \tag{5.51}$$

[a]Asymptotic t-ratio.
[b]Partial elasticity evaluated at sample means.

The planning horizon of pork producers is assumed to be two quarters (h = 2 quarters). For each quarter, the model was solved ahead for eight quarters (k = 6). The convergence criterion for the set of Type I Gauss-Seidel iterations was 10^{-5}, and the convergence criterion for the Type II iterations was 10^{-4}. The overall convergence criterion was equal to 10^{-6}.

The starting values for the solution and subsequent estimation were obtained by replacing the rational expectation of the farm price of barrows and gilts ($_tFPPK_{t+2}$) in (5.37) and (5.38) with a two-step-ahead ARIMA forecast. The solution steps were omitted because the rational expectation was replaced by the ARIMA projection. The starting values were set to the FIML estimates. The ARIMA model used to generate the forecasts of the farm price of barrows and gilts was

$$(1 - B)FPPK_t = (1 - 0.5807B^5) \; \epsilon_t \; ,$$

$$(-5.32)$$

(5.52)

$$Q(18) = 20.05 \qquad SE = 4.59$$

where the 0.05 critical value of the x^2 distributed Box-Pierce Q statistic with 18 degrees of freedom is 28.87. The components of the ARIMA model have the same interpretation as in Table 5.2.

The FIML estimates, obtained from replacing the rational expectation with the ARIMA forecast, were used as the starting values in the solution method. The solution procedure obtained the rational expectation of the farm price of barrows and gilts. The rational expectation of the farm price of barrows and gilts was used in the subsequent FIML estimation of the model. This procedure of obtaining starting values greatly reduced the need for successive solution-estimation iterations. The rational expectation estimates were quite similar to the ARIMA forecast estimates. This would imply that much of the information in the model is already captured in the series of past prices of barrows and gilts.

Estimation required 3,685 evaluations of the log-likelihood function, which achieved a maximum value of -887.45. The total CPU time required to obtain the rational expectation solution and for estimation was 44.03 minutes on an IBM 9377 minicomputer. The FIML estimation was completed with the DFP subroutine in GQOPT (Quandt and Goldfeld 1987). Although time-consuming, the resulting parameter estimates were generally significant at conventional levels and all were of anticipated sign.

The additions to the breeding herd (5.37) were more responsive to anticipated output and input prices than the level of sow slaughter (5.38). This result is fairly intuitive because a large segment of sow slaughter is due to age, not economic rationale.

The liveweights of barrows and gilts (5.43) and sows (5.44) were both responsive to contemporaneous changes in output and input prices. Also, a positive upward trend was found in both liveweights. However, the trend in the liveweight of barrows and gilts, while positive, is insignificant at conventional levels.

In the retail demand equation (5.49), a zero-one dummy variable (D794) was included after initial estimates were obtained. The zero-one dummy variable is equal to one starting in the fourth quarter of 1979 and is equal to zero before that date. The dummy variable was included because the retail demand equation was overpredicting the retail price of pork during the 1980s. The negative sign on D794 suggests a downward demand shift during that period. Reasons for the apparent demand shift include changes in tastes and preferences due to health concerns and changes in the macroeconomic environment. The time trend in the retail demand was negative, but insignificant.

Thus, evidence of a continual, downward demand shift over the sample period is lacking.

The estimation results of the retail-farm margin (5.50) suggest that the third and fourth quarters are slightly more responsive to changes in the retail price of pork. Also, the results indicate that, as the total retail value of production increases, holding population constant, the retail-farm margin increases. The index of packer marketing costs proved to be a positive, yet insignificant, factor in determining the retail-farm margin. The magnitude and sign of the lagged dependent variable suggest that a good degree of stickiness is found in the transmission of prices from the retail to farm level.

Futures Market Expectation Model

Futures market prices provide an alternative form of price expectations. With futures market expectations, the price determination is removed from the model structure. Thus, the demand component of the model is unneeded. Also, only the supply components that determine the supply categories of interest are needed, namely additions to the breeding herd, sow slaughter, breeding herd inventory, sows farrowing, and the pig crop. Of these, additions to the breeding herd and the sow slaughter, as specified, are the only categories influenced by changes in economic factors. These two equations remain restricted by biological restrictions. Additions to the breeding herd less sow slaughter, plus the level of breeding herd inventory in the previous quarter determine the current period's breeding herd inventory through an identity. The remaining supply equations for sows farrowing and the pig crop continue to be restricted by biological restrictions and are the same as reported in the supply structure of the rational expectation model [Table 5.3, Equations (5.40) and (5.41)].

Quarterly averages of closing futures market prices for live hogs and corn are used as profitability expectations in additions to the breeding herd and sow slaughter equation. The soybean meal futures price was excluded to ease the burden of data collection. The expectations are formed when the sow or gilt is bred and are for the live hog and corn contract nine to ten months in the future (see Table 4.2). The supply components with futures market expectations were estimated from 1970 to 1986 with restricted least squares (RLS).

As provided in Table 5.4, the additions to the breeding herd (5.53) with futures market expectations are more responsive to economic variables than in the rational expectation model. This holds true for the sow slaughter (5.54) equation as well. The parameter estimates have the anticipated signs and are significant at conventional levels. The results indicate that futures market prices for live hogs (FUTHOG) and corn (FUTCORN) provide a viable alternative to capturing the breeding herd decisions of hog producers.

Model Validation

The significance of the parameter estimates for the rational expectation model (Table 5.3) and for the futures market expectation model (Table 5.4) indicate the goodness-of-fit of these two alternative representations of the pork sector. However, the adequacy of the models is better judged in a system context. Two criteria are used to assess the validity of the models: historical simulation and comparison with previous econometric models of the pork sector.

Historical simulation evaluates how well the estimated model tracks the underlying historical data series. The rational expectation and futures market expectation models were simulated during the sample period, 1970-86. Several simulation statistics are given for selected equations in the rational

Table 5.4. Estimation results of the futures market expectation model, 1970-86

Additions to the Breeding Herd

$ABHUS_t$ = 0.0434 * D1 * $PCUS_{t-2}$ + 0.0459 * D2 * $PCUS_{t-2}$

$\qquad$ + (0.0773 + 0.1457 * D747 - 0.0164 * DL74 * T65) * D3 * $PCUS_{t-2}$

$\qquad$ + (0.0376 - 0.0443 * DL74 + 0.00474 * DL74 * T65) * D4 * $PCUS_{t-2}$

$\qquad$ + 23.913 * $FUTHOG_t$ - 348.966 * $FUTCORN_t$
$\qquad\quad$ (3.29) $\qquad\qquad\qquad$ (-3.09)
$\qquad\quad$ [0.85] $\qquad\qquad\qquad$ [-0.78]

$\qquad$ - 22.465 * $RIFCL_t$
$\qquad\quad$ (-2.68)
$\qquad\quad$ [-0.09]

$\qquad\quad$ R^2 = 0.93[a] $\qquad\qquad\qquad$ D.W. = 1.95[b] $\qquad\qquad\qquad\qquad$ (5.53)

Sow Slaughter

$SSUS_t$ = (0.1183 - 0.3599 * DUM76 + 0.00237 * DUM76 * T65) * D1 * $BHUS_{t-1}$

$\qquad$ + (0.1312 - 0.0694 * DUM76 + 0.00348 * DUM76 * T65) * D2 * $BHUS_{t-1}$

$\qquad$ + (0.1620 - 0.0815 * DUM76 + 0.00394 * DUM76 * T65) * D3 * BHUS

$\qquad$ + (0.1543 - 0.0393 * DUM76 + 0.00169 * DUM76 * T65) * D4 * $BHUS_{t-1}$

$\qquad$ - 5.831 * $FUTHOG_t$ + 94.419 * $FUTCORN_t$ + 0.2007 * $RIFCL_t$
$\qquad\quad$ (-2.32) $\qquad\qquad$ (2.42) $\qquad\qquad\qquad$ (0.07)
$\qquad\quad$ [-0.20] $\qquad\qquad$ [0.21] $\qquad\qquad\qquad$ [0.0008]

$\qquad\quad$ R^2 = 0.99 $\qquad$ D.W. = 1.13 $\qquad\qquad\qquad\qquad\qquad$ (5.54)

[a]R^2 is the squared correlation coefficient.
[b]D.W. is the Durbin-Watson d Statistic.

expectation model and in the futures market expectation model. The historical simulation statistics for the rational and futures market expectations are provided in Tables 5.5 and 5.6, respectively. The simulation statistics provided Theil's inequality coefficient ($0 < U < 1$), the root mean square error (RMSE), and the root mean percent square error (RMPSE).

Theil's inequality coefficient can be decomposed into these proportions of inequality: $U^M + U^S + U^C = 1$ (Pindyck and Rubinfeld 1981). The three proportions U^M, U^S, and U^C are called the bias, the variance, and the covariance proportions, respectively. The bias measures the systematic error in simulation. The variance indicates how well the model replicates the degree of variability in the historical series. The covariance measures the unsystematic error. It is desirable to have small values of U^M and U^S relative to U^C.

The RMSE measures the deviation of the simulated variable from its historical values. The magnitude of the RMSE is judged relative to the size of values in the observed historical series. The RMPSE measures the deviation of the simulated variable from the historical values in percentage terms.

The simulation statistics for the rational expectation model are generally good. However, the bias component for the liveweights and the retail and farm prices are higher than desired. Also, the other simulation statistics indicate that price determinations of the model could be improved. Furthermore, additions to the breeding herd equation do not adequately capture the breeding herd investment decisions of producers. In part, this is caused by the source of the data; it is created as a residual of the breeding herd inventory plus sow slaughter. Thus, the series includes measurement errors from the breeding herd inventory and sow slaughter data.

Table 5.5. Rational expectation model simulation statistics, 1970-86

Variable	Label	Measure[a]				
		U^M	U^S	U^C	RMSE	RMPSE
Additions to the breed herd (5.37)	ABHUS	0.000	0.070	0.930	387.14	69.59
Sow slaughter (5.38)	SSUS	0.078	0.000	0.922	170.99	14.93
Breeding herd inventory (5.39)	BHUS	0.014	0.029	0.956	422.67	5.12
Sows farrowing (5.40)	FARROW	0.010	0.053	0.937	191.50	6.78
Pig crop (5.41)	PCUS	0.007	0.022	0.971	1519.48	6.71
Barrow and gilt slaughter (5.42)	BGSUS	0.000	0.001	0.999	618.22	3.29
Liveweight of barrows and gilts (5.43)	LWBG	0.339	0.106	0.555	10.55	2.32
Liveweight of sows (5.44)	LWS	0.371	0.015	0.614	7.80	3.27
Commercial pork production (5.46)	TOTSPK	0.098	0.211	0.691	156.91	4.32
Retail pork price (5.49)	RPPK	0.325	0.222	0.452	0.05	8.87
Retail-farm margin (5.50)	MARGIN	0.026	0.272	0.703	0.02	4.34
Farm price of barrows and gilts (5.51)	FPPK	0.518	0.027	0.456	11.78	25.11

[a]The U values are the components of Theil's inequality coefficient. U^M is the bias component, U^S is the variance component, and U^C is the unsystematic component. RMSE is the root mean square error and RMPSE is the root mean percent square error.

Table 5.6. Futures market expectation model simulation statistics, 1970-86

Variable	Label	Measure[a]				
		U^M	U^S	U^C	RMSE	RMPSE
Additions to the breed herd (5.53)	ABHUS	0.001	0.096	0.903	307.69	56.33
Sow slaughter (5.54)	SSUS	0.143	0.002	0.855	172.44	15.47
Breeding herd inventory (5.39)	BHUS	0.052	0.002	0.946	331.57	3.88
Sow farrowing (5.40)	FARROW	0.034	0.016	0.950	150.04	4.64
Pig crop (5.41)	PCUS	0.022	0.001	0.976	1249.23	5.33

[a]The U values are the decomposed components of Theil's inequality coefficient. U^M is the bias component, U^S is the variance component, and U^C is the unsystematic component. RMSE is the root mean square error and RMPSE is the root mean percent square error.

The simulation statistics for the futures market model are in general better than the rational expectation model. Eliminating price determination from the model structure reduces a source of error. This, in turn, improves the model's ability to track the historical data series.

The elasticity of supply indicates behavior of econometric models. The supply elasticities for selected econometric models of the pork sector are provided in Table 5.7. The supply elasticity of the rational expectation model is generally lower than the previous results. The futures market expectation model closely resembles other models' supply response behavior. Of course, there are many reasons for the differences among the estimated supply elasticities. The period of study is one reason. Differences in the method of calculation of the elasticity also can affect its value.

Table 5.7. Comparison of selected pork supply response elasticities

Study	Data	Period	Supply Elasticity	
Dean and Heady (1958)	Semiannual	1938-56	"Spring	0.60
			Fall"	0.30
		1924-37	"Spring	0.50
			Fall"	0.28
Cromarty (1959)	Annual	1929-53		0.13
Harlow (1962)	Annual	1949-60		0.56 to 0.82
Meilke, Zwart, and Martin (1974)	Quarterly	1961-71		0.43 to 0.48
Heien (1975)	Annual	1950-69		0.31
Marsh (1977)	Annual	1953-75		0.36
MacAulay (1978)	Quarterly	1966-76		0.50
Skold and Holt (1988)	Quarterly	1967-85		0.23
Skold, Grundmeier, and Johnson (1988)	Quarterly	1967-86		0.03[a] 0.50
Rational Expectation	Quarterly	1970-86		0.02[a] 0.27
Futures Market Expectation	Quarterly	1970-86		0.08[a] 0.43

[a]Denotes short-run elasticity is generally lower than previous results.

Summary

In this study, supply elasticities were estimated through simulation in the spirit of Fair (1980). In brief, the exogenous variables were set to their sample means. The models were then simulated until steady-state solutions were obtained. This solution formed a baseline for comparison. Then, the models were simulated again, and the output price was perturbed for four quarters. The average impact of the output price change on commercial supply of pork in the first year is defined as the short-run elasticity. The long-run elasticity was measured at the point of a new steady-state solution after the shock in the output prices. The new steady-state solution was obtained within twelve quarters.

Chapter 6

Composite Prediction
of the *Hogs and Pigs* Report

The initial estimates of supply and inventory categories in the *Hogs and Pigs* report continue to rely primarily on the quarterly surveys of hog producers. The judgment of National Agricultural Statistical Service (NASS) officials in the data evaluation process is an additional factor in the development of the estimates. Other sets of information beyond the survey data can be incorporated into the construction of the initial estimates so that estimate reliability and precision are improved. Specifically, the forecasts from the rational and futures market expectation models can be incorporated into the data evaluation process through composite forecasting techniques.

In general, by combining different sets of information into a single composite forecast, an individual prediction is obtained that is more accurate than any of its individual components (Johnson and Rausser 1982). This desirable property has motivated the use of composite forecasting in providing outlook projections of prices and supplies in often unstable agricultural markets. For example, Brandt and Bessler (1981; 1983), and Bessler and Brandt (1979) combined forecasts from econometric and ARIMA models with projections of experts to obtain hog price forecasts. Shideed and White (1988) applied various composite forecasting techniques to corn and soybean meal cash and futures prices to improve estimates of U.S. soybean acreage. Finally Krog (1988) used composite methods to combine NASS corn yield estimates with yield estimates from a plant process model.

In this chapter, the USDA initial estimates of the U.S. breeding herd, sows farrowing, and the pig crop are combined with forecasts from the RE and FME models to form a single prediction for these respective supply and inventory categories. Further discussion of composite forecasting methods can be found in Johnson and Rausser (1982), Clemen and Winkler (1986), and Granger and Newbold (1986).

Both nonstochastic and stochastic composite forecasting techniques are used in the current study. The nonstochastic methods are based on the estimated variance-covariance matrix of the forecast errors from the RE and FME models and the USDA initial estimates. The nonstochastic methods assume the USDA final estimate is not a random variable and allow for the possibility of biases in the market model forecasts and the USDA initial estimates. The nonstochastic methods are extended to allow for intertemporal variation in the weights associated with the three forecasts.

The stochastic methods include estimating the fixed composite forecasts with a random disturbance term. Thus, with the stochastic methods, the USDA final estimates are assumed to be random variables. The other stochastic methods used are OLS, with and without linear parameter constraints, and ridge regression.

The Individual Forecasts

The quarterly forecasts from the RE model are one-step-ahead forecasts of the U.S. breeding herd inventory (REBH), sows farrowing (REFAR), and the pig crop (REPC). The RE forecasts for these variables are conditioned on the ARIMA projections of stochastic exogenous variables and the RE forecasts of additions to the breeding herd, sow slaughter, and the farm price of barrows and gilts. The stochastic exogenous variables generated by the

ARIMA forecasts include predictions of the feed cost of the index, interest rates, food expenditures, and the real retail price of beef (Table 5.2).

The FME forecasts for the U.S. breeding herd inventory (FUTBH), sows farrowing (FUTFAR), and the pig crop (FUTPC) are conditioned on the expected spot prices of live hogs and corn, represented by the quarterly average of the closing contract prices nine to ten months in the future. The FME estimates also are conditioned on interest rate projections. The forecast of the interest rate was assumed to be the current period's rate. Thus, futures market participants are assumed to use a naive forecasting mechanism to project this exogenous variable.

The USDA initial estimates for the U.S. breeding herd inventory (USDABH), sows farrowing (USDAFAR), and the pig crop (USDAPC) are used as the USDA forecasts. At the U.S. aggregate level, however, initial estimates are only available for the breeding herd inventory and pig crop in the December and June reports. This required interpolations with the ten-state initial estimates that are available quarterly. The interpolations assumed that the percentage change in these initial estimates at the aggregate U.S. level would be the same as the quarter-to-quarter percentage change in the ten-state initial estimates (see Appendix A for details).

Some further interpolations were necessary for the pig crop. The initial U.S. estimates in the December and June reports up until 1978 were reported as estimates of the pig crop born in the previous six months. For the USDA initial pig crop estimates (USDAPC), the distribution of pig births between the previous two quarters was assumed to follow the distribution implied by the finalized ten-state estimates. Initially, the quarter-to-quarter distribution was assumed to follow the implied birth distribution indicated by the initial ten-state estimates. This resulted in large forecasting errors in the interpolated quarters; therefore, the ten-state final estimates were used for the interpolations. Beginning in 1978, initial estimates of the U.S. pig crop became available on a quarterly basis. These interpolations, of course, may add forecast error to the initial USDA estimates. The amount of error depends on the forecast error in the ten-state initial estimates and on the differences in the movements in the U.S. supply and inventory categories relative to the ten-state region.

The quarterly forecasts for the U.S. breeding herd inventory, sows farrowing, and the pig crop were made for 1970 to 1986. This represented 68 forecasts for these supply and inventory variables. The USDA final estimates for the U.S. breeding herd inventory (BHUS), sows farrowing (FARROW), and the pig crop (PCUS) are assumed to be the actual values and are available through 1982. After 1982, the final estimates were obtained from the latest *Hogs and Pigs* report (USDA 1970-88), which estimates the current period and the previous two years. The estimates can be revised with each subsequent report. The 1982-86 initial estimates were not finalized until after the release of the 1987 Census of Agriculture.

Forecast Accuracy

The selected forecast statistics used to quantify the accuracy of the RE and FME model forecasts and the USDA initial estimates, as well as subsequent composite forecasts, included the root mean square error (RMSE). The RMSE is defined as

$$RMSE = \left[\frac{1}{T} \sum_{t=1}^{T} (Y_{at} - Y_{pt})^2 \right]^{1/2} , \qquad (6.1)$$

where Y_{at} and Y_{pt} are the actual and predicted values, respectively, for the t = 1, ..., T. RMSE indicates the average deviation of the forecast from the actual value. This measure in percentage terms is the root mean percent square error (RMPSE) and is defined as

$$RMPSE = \left[\frac{1}{T} \sum_{t=1}^{T} \left(\frac{Y_{at} - Y_{pt}}{Y_{at}} \right)^2 \right]^{1/2} . \tag{6.2}$$

Another measure is the mean absolute error (MAE). The MAE penalizes large errors less than the RMSE. The MAE is defined as

$$MAE = \frac{1}{T} \sum_{t=1}^{T} |Y_{at} - Y_{pt}| . \tag{6.3}$$

All of these forecast accuracy measures are equal to zero if the forecasts are perfect.

The ratio of the mean square error (MSE) for the given forecast (e.g., rational, composite) to the USDA initial estimate MSE also is provided. This also indicates relative forecast efficiency.

Individual Forecast Accuracy

As expected, the USDA initial estimates were superior to the RE and FME forecasts. These initial estimates contain more recent information about the activities of hog producers, and should provide a better estimate of supply and inventory categories. The forecast statistics for the RE and FME models and for the initial USDA estimates are provided in Table 6.1. In Figures 6.1, 6.2, and 6.3, the percentage forecast errors of the breeding herd inventory are presented for the RE, FME, and USDA initial estimate predictions, respectively. The sows farrowing forecast errors for the RE, FME, and

Table 6.1. Individual estimates forecast accuracy, 1970-86

Variable (Label)	Measures[a]			
	RMSE	RMPSE	MAE	MSE ratio
Breeding Herd Inventory				
USDA (USDABH)	103.4	1.19	71.1	
Rational expectation (REBH)	296.2	3.64	233.8	8.21
Futures-Market Expectation (FUTBH)	294.5	3.50	242.4	8.11
Sows Farrowing				
USDA (USDAFAR)	64.3	2.27	36.2	
Rational expectation (REFAR)	154.8	5.02	112.4	5.79
Futures market expectation (FUTFAR)	153.2	4.08	119.2	5.67
Pig Crop				
USDA (USDAPC)	382.0	1.77	236.7	
Rational expectation (REPC)	1229.5	5.40	909.9	10.36
Futures market expectation (FUTPC)	1244.7	5.28	968.1	10.62

[a]RMSE is the root mean square (6.1); RMPSE is the root mean percent square error (6.2); MAE is the mean absolute error (6.3); and MSE ratio is the ratio of the MSE to the MSE/USDA initial estimates.

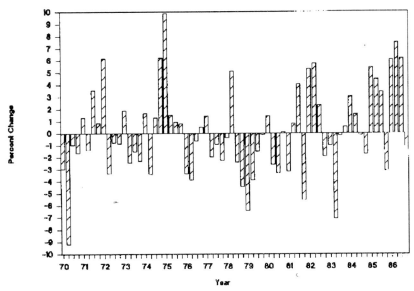

Figure 6.1. Percentage change between rational expectation forecast and the USDA final estimate: U.S. hogs kept for breeding, 1970-286

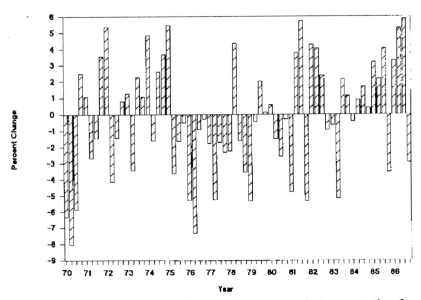

Figure 6.2. Percentage change between futures market expectation forecast and the USDA final estimate: U.S. hogs kept for breeding, 1970-86

USDA initial estimates are given in Figures 6.4, 6.5, and 6.6, respectively. Similarly, the pig crop forecast errors are given in Figures 6.7, 6.8, and 6.9.

On the basis of an MSE criterion, the FME model provided slightly better forecasts than the RE model for the breeding herd inventory and for sows farrowing, the RE estimates of the pig crop were better than the FME estimates, and both RE and FME models were clearly inferior to USDA predictions. However, this does not preclude gleaning information from the RE and FME estimates.

Combining Forecasts

The forecasts of the RE and FME models and USDA initial estimates use different sets of information. Consequently, this implies that individual point forecasts from each forecasting system will be different and, in general, not optimal. Composite forecasting provides an eclectic approach that balances each system's individual forecast into a single composite prediction. The methods in this study provide different approaches to estimating the appropriate weights on the individual forecasts in forming the composite prediction. Again, regardless of the estimation technique used, the rationale is transparent—the single composite prediction will, in general, outperform the individual components.

Bates and Granger (1969) provide an early example of composite forecasting techniques. They combined two unbiased predictions of world airline passenger estimates into a single prediction. Their methods provide background and motivation for other composite forecasting methods.

Let Y_n^1 and Y_n^2 be two unbiased forecasts of y_n for n periods. The forecast error in period n is defined as

$$e_n^i = y_n - Y_n^i, \quad i = 1, 2, \tag{6.4}$$

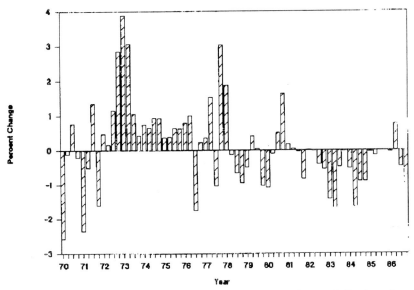

Figure 6.3. Percentage change between USDA initial and final estimates: U.S. hogs kept for breeding, 1970-86

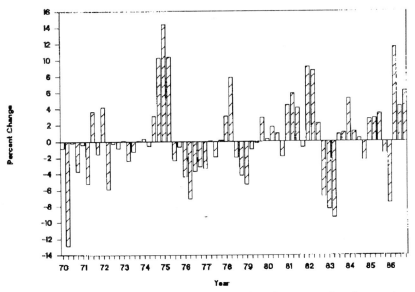

Figure 6.4. Percentage change between rational expectation forecast and the USDA final estimate: U.S. sows farrowing, 1970-86

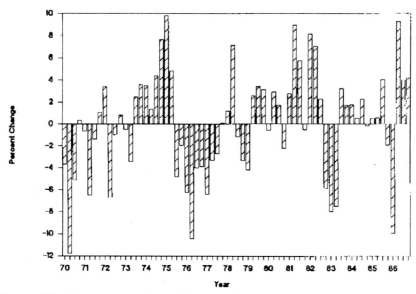

Figure 6.5. Percentage change between futures market expectation forecast and the USDA final estimate: U.S. sows farrowing, 1970-86

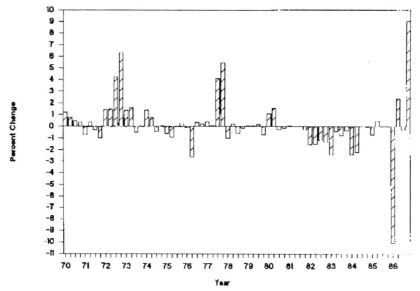

Figure 6.6. Percentage change between USDA initial and final estimates: U.S. sows farrowing, 1970-86

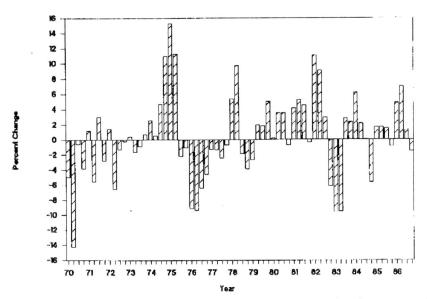

Figure 6.7. Percentage change between rational expectation forecast and the USDA final estimate: U.S. pig crop, 1970-86

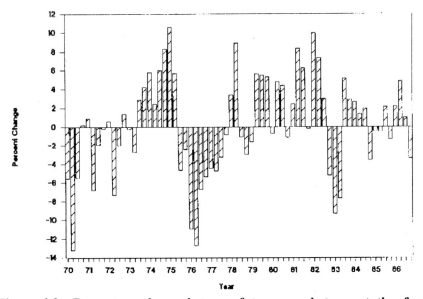

Figure 6.8. Percentage change between futures market expectation forecast and the USDA final estimate: U.S. pig crop, 1970-86

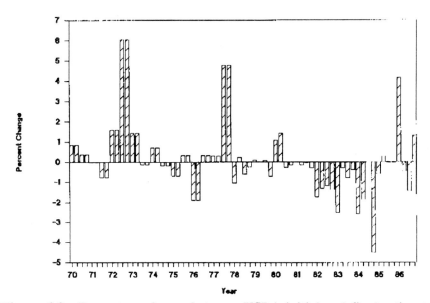

Figure 6.9. **Percentage change between USDA initial and final estimates: U.S. pig crop, 1970-86**

with

$$E(e_n^i) = 0, \ E(e_n^{i2}) = \sigma_i^2, \quad i = 1, 2, \tag{6.5}$$

and

$$E(e_n^1 \ e_n^2) = \rho \sigma_1 \ \sigma_2, \tag{6.6}$$

where ρ is the correlation coefficient of the two forecast errors, e_n^1 and e_n^2.

The forecasts are combined as a weighted average of the two individual forecasts,

$$C_n = kY_n^1 + (1 - k) \ Y_n^2, \tag{6.7}$$

where C_n is the composite forecast and k is the weight. The composite forecast error is

$$e_n^c = ke_n^1 + (1 - k) \ e_n^2, \tag{6.8}$$

with error variance

$$\sigma_c^2 = k^2\sigma_1^2 + (1 - k) \ \sigma_2^2 + 2k \ (1 - k) \ \rho\sigma_1 \ \sigma_2. \tag{6.9}$$

Bates and Granger show that the composite error variance σ_c^2 is minimized for the value of k equal to

$$k_o = \frac{\sigma_2^2 - \rho \ \sigma_1 \ \sigma_2}{\sigma_1^2 + \sigma_2^2 + 2\rho\sigma_1 \ \sigma_2}. \tag{6.10}$$

Substituting this optimal k_o into (6.9) yields the minimum achievable forecast error variance,

$$\sigma_{c,o}^2 = \frac{\sigma_1^2 \ \sigma_2^2 \ (1 - \rho^2)}{\sigma_1^2 + \sigma_2^2 - 2\rho\sigma_1^2}. \tag{6.11}$$

The forecast error variance $\sigma^2_{c,o}$ is always at least as small as the minimum error variances of the individual forecasts σ^2_1 and σ^2_2. If ρ is equal to σ_1/σ_2 or σ_2/σ_1, the variance of the combined forecast equals the smaller of the two individual error variances.

Bates and Granger developed methods to obtain the maximum likelihood estimate of k_o by assuming that the individual forecast errors are distributed bivariate normal. Also, they considered alternative estimators for k_o that allowed changes in the weights across time.

Nonstochastic Methods

The methods developed by Bates and Granger (1969) estimate the optimal weights by analytical methods and do not assume that other random factors influence the determination of the true value being predicted. Thus, with nonstochastic methods, the USDA final estimate is assumed to be fixed. Bates and Granger's estimation methods have been extended to the example of more than two competing forecasts by Newbold and Granger (1974). As with the two competing forecasts, the estimation methods for more than two forecasts include estimators that permit changes in the weights through time and incorporate the correlation between the individual forecasts.

The nonstochastic methods applied to the competing forecasts of the breeding herd inventory, sows farrowing, and the pig crop rely on different sorts of estimates of the sample error variance-covariance matrix. The first two nonstochastic methods assume fixed weights on the three competing forecasts. The second two nonstochastic methods allow intertemporal changes in the weights.

Estimated Variance-Covariance

The first nonstocastic method applied to the competing forecasts used a sample estimate of the error variance-covariance matrix. This method, developed by Newbold and Granger (1974), is presented in Granger and Newbold (1986).

Using similar notation, let $Y_n' = Y_n^1, ..., Y_n^m)$ be the vector of m competing forecast for n time periods. The individual forecast error is defined as $e_n = y 1 - y_n$, with $E(e_n e_n') = \Sigma$ and $1' = (1, ..., 1)$. As before, the actual value is defined as y. The composite forecast is defined as a weighted average of the individual estimates

$$c_n = k_n' Y_n, \quad k_n' \ 1 = 1, \tag{6.12}$$

such that $0 \leq k_n^i \leq 1$, $i = 1, ..., m$, and where $k_n' = (k_n^1, ..., k_n^m)$, the weights on the individual forecasts. The composite forecast error is minimized with weights equal to

$$k_o = (\textstyle\sum^{-1} 1)/(1' \textstyle\sum^{-1} 1) . \tag{6.13}$$

Breeding Herd Inventory. After obtaining an estimate of the error variance-covariance matrix $\hat{\Sigma}_{BH}$ of the forecast errors from the RE, FME, and USDA initial estimates, the weight estimates are equal to

$$\hat{k}_o' \ (\hat{\Sigma}_{BH}) = \begin{bmatrix} -0.0072 \\ 0.1173 \\ 0.8898 \end{bmatrix}, \tag{6.14}$$

where $\hat{k}_o' \ (\hat{\Sigma}_{FAR}) = $ (REBH, FUTBH, USDABH). The weights do sum to one but were not constrained to be positive. Consequently, the weight on the RE estimate is slightly negative. In general, this is not a desired result because it implies that the RE model provides an inferior forecast. The RE negative weight estimate was retained in the composite forecast on the grounds that the relatively greater error variance is outweighed by the relatively high correlation among the forecasts [Johnson and Rausser (1982)]. That is, part of the actual breeding herd inventory left unexplained by the RE estimate is sufficiently strongly related to the part unexplained by the FME and USDA forecasts.

Sows Farrowing. Similarly, the weight estimates for the sows, obtained from the error variance-covariance matrix $\hat{\Sigma}_{FAR}$, are

$$\hat{K}_o(\hat{\Sigma}_{FAR}) = \begin{bmatrix} 0.1163 \\ 0.0058 \\ 0.8778 \end{bmatrix}, \tag{6.15}$$

where $\hat{k}_o{}'(\hat{\Sigma}_{FAR}) = $ (REFAR, FUTFAR, USDAFAR). The weight on the RE forecast is positive, unlike the breeding herd inventory composite estimate. The largest weight is on the USDA initial estimate. The RE weight is greater than the FME weight.

Pig Crop. The weight estimates for the pig crop are similar to the sows farrowing results. The RE estimate receives greater weight than the FME estimate, and the USDA estimate is weighted most heavily. The estimated weights for the pig crop are

$$\hat{k}_o(\hat{\Sigma}_{pc}) = \begin{bmatrix} 0.0658 \\ 0.0365 \\ 0.8976 \end{bmatrix}, \tag{6.16}$$

where $\hat{k}_o{}'(\hat{\Sigma}_{pc}) = $ (REPC, FUTPC, USDAPC), conditioned on the estimate of the error variance-covariance matrix $\hat{\Sigma}_{pc}$.

Estimated Variance-Covariance Matrix with Bias

This second nonstochastic estimation method is similar to the above method, but it allows for bias in the RE, FME, and USDA estimates of the supply and inventory categories. The estimate of the error variance-covariance matrix is not computed directly. Rather, it is obtained from some preliminary OLS regressions.

The concept underlying this estimator is closely related to the inverse regression problem (Draper and Smith 1981, 47-51). Consider two estimates that are available for some phenomenon, Y and X. Assume that X provides the more sure estimate. The fitted regression equation $Y_i = \beta_0 + \beta_1 X_i$ provides a "calibration curve" for the less precise variable Y related to the more precise measure X. Given an estimate of Y, say Y_0, the predicted value of the actual value is

$$X_o = \beta_1^{-1} (Y_o - \beta_o) . \qquad (6.17)$$

Thus, in applying this technique to the composite forecasting estimator, the RE, FME, and initial USDA estimates were regressed on the actual values. To relate to the example, Y is the RE, FME, and USDA initial estimates, and X is the final USDA estimate. The estimation results for these OLS calibration regressions are provided in Table 6.2.

The residuals from the fitted equations in Table 6.2 are used to estimate an error variance-covariance matrix that incorporates the slope biases in the estimates. This estimate of the variance-covariance matrix V is of the general form

Table 6.2. OLS calibration estimation results of individual forecasts on final
USDA estimates

Dependent Variable	Intercept	Slope	Explanatory Variable	R^{2a}	D.W.[b]	Equation
USDABH	-43.95 (-.43)[c]	1.007 (82.5)	BHUS	0.99	1.08	(6.19)
REBH	1072.10 (4.08)	0.87 (27.8)		0.92	1.72	(6.20)
FUTBH	686.96 (2.47)	0.92 (27.6)		0.92	1.59	(6.21)
USDAFAR	-21.94 (0.39)	1.009 (56.1)	FARROW	0.98	1.45	(6.22)
REFAR	491.55 (4.07)	0.84 (21.6)		0.87	1.31	(6.23)
FUTFAR	327.86 (2.58)	0.89 (21.8)		0.88	1.12	(6.24)
USDAPC	-120.49 (-0.37)	1.0007 (69.8)	PCUS	0.99	1.05	(6.25)
REPC	3145.9 (3.19)	0.86 (19.8)		0.86	1.02	(6.26)
FUTPC	2028.5 (1.95)	0.91 (19.8)		0.86	0.87	(6.27)

[a]R^2 is the multiple correlation coefficient.
[b]D.W. is the Durbin-Watson d statistic.
[c]The t-ratio for the estimated coefficient is in parentheses.

$$
V = \begin{bmatrix} \begin{pmatrix} \beta_{11}^{-1} & 0 & 0 \\ 0 & \beta_{12}^{-1} & 0 \\ 0 & 0 & \beta_{13}^{-1} \end{pmatrix} S_{\epsilon\epsilon} \begin{pmatrix} \beta_{11}^{-1} & 0 & 0 \\ 0 & \beta_{12}^{-1} & 0 \\ 0 & 0 & \beta_{13}^{-1} \end{pmatrix} \end{bmatrix}, \tag{6.18}
$$

where $S_{\epsilon\epsilon}$ is the variance-covariance matrix of the residuals from the calibrated equations (Table 6.2), and β_{11}, β_{12}, β_{13} are estimates of slope parameters. The estimates of the weights $k_b(V)$ are obtained by replacing Σ^{-1} by V^{-1} in (6.14). The composite forecast C_n is then obtained by

$$
C_n = k_b(V) \begin{bmatrix} \beta_{11}^{-1} \ (RM - \beta_{01}) \\ \beta_{12}^{-1} \ (FME - \beta_{02}) \\ \beta_{13}^{-1} \ (USDA - \beta_{03}) \end{bmatrix}, \tag{6.28}
$$

where RE, FME, and USDA are the individual forecasts, and β_{01}, β_{02}, and β_{03} are the estimated intercept parameters from Table 6.2.

Breeding Herd Inventory. By using the residuals from equations (6.20) through (6.22) as listed in Table 6.2, the estimate of the variance-covariance matrix $\hat{V}_{BH}$ was formed. From this the estimates of k are then

$$
\hat{k}_b(\hat{V}_{BH}) = \begin{bmatrix} -0.0175 \\ 0.1083 \\ 0.9092 \end{bmatrix}, \tag{6.29}
$$

where $\hat{k}_b'(\hat{V}_{BH}) = [\hat{\beta}_{11}^{-1}(REBH - \hat{\beta}_{01}), \hat{\beta}_{12}^{-1} (FUTBH - \hat{\beta}_{02}), \hat{\beta}_{13}^{-1} (USDABH - \hat{\beta}_{03})]$. Similar to (6.18), the parameters are the intercept and slope coefficients from equations (6.19) through (6.21). The weight estimates are quite similar to those obtained earlier in (6.14). However, the weights on the RE and FME estimates are less, and the USDA initial estimate has a greater weight in the composite forecast.

Sows Farrowing. Again, the residuals from the fitted equations (6.22) through (6.24) and the slope parameters were used to estimate the variance-covariance matrix $\hat{V}_{FAR}$. The weight estimates are

$$\hat{K}'_b \, (\hat{V}_{FAR}) = \begin{bmatrix} 0.0816 \\ 0.0113 \\ 0.9071 \end{bmatrix} \qquad (6.30)$$

where $\hat{k}_b{}'(\hat{V}_{FAR}) = [\hat{\beta}_{11}{}^{-1}(REFAR - \hat{\beta}_{01}), \hat{\beta}_{12}{}^{-1} (FUTFAR - \hat{\beta}_{02}), \hat{\beta}_{13}{}^{-1}$ $(USDAFAR - \hat{\beta}_{03})]$. Allowing for biases in the individual forecasts increases the weight on the USDA initial estimate.

Pig Crop. Similarly, the residuals from the fitted equations (6.25)-(6.27) and slope coefficients formed the estimate of the error variance-covariance matrix for the pig crop estimates $\hat{V}_{PC}$. The resulting weight estimates are

$$\hat{k}_b(\hat{V}_{PC}) = \begin{bmatrix} 0.0457 \\ 0.0396 \\ 0.9146 \end{bmatrix} , \qquad (6.31)$$

where $\hat{k}_b{}'(\hat{V}_{PC}) = [\hat{\beta}_{11}{}^{-1}(REPC - \hat{\beta}_{01}), \hat{\beta}_{12}{}^{-1} (FUTPC - \hat{\beta}_{02}), \hat{\beta}_{13}{}^{-1}$ $(USDAPC - \hat{\beta}_{03})]$. Compared to the pig crop estimates in (6.16), the FME and USDA weights slightly increase, and consequently the RE weight decreased.

Time-Varying Weights

The weights on the individual forecasts may change during the sample period. Changes in the weights may be due to improved USDA sampling and survey procedures and shifts in NASS personnel. Also, one forecasting procedure may provide preferred estimates at certain times of the hog cycle. For example, the USDA initial estimates may miss turning points in the hog cycle, whereas the RE estimates, although inferior, accurately predict the turn in the breeding herd inventory.

Two different estimators were used that adapted quickly to the relative magnitude of forecast variances among the competing predictions. The two estimators are

$$\hat{k}_n^i = (\sum_{t=n-v}^{n-1} e_t^{i2})^{-1} / (\sum_{j=1}^{m} (\sum_{t=n-v}^{n-1} e_t^{j2})^{-1}), \qquad (6.32)$$

and

$$\hat{k}_n^i = \alpha \hat{k}_{n-1}^i + (1 - \alpha)(\sum_{t=n-v}^{n-1} e_t^{i^2})^{-1} / (\sum_{j=1}^{m} (\sum_{t=n-v}^{n-1} e_t^{j2})^{-1}), \qquad (6.33)$$

where $\hat{k}_n^i$ is the weight on the ith forecast in period n, M is the number of individual forecasts, and α and v are chosen arbitrarily (Granger and Newbold 1986).

These time-varying weight estimators were chosen because of the previous successes of Newbold and Granger (1974) and Winkler and Makridakis (1983). Their results suggest that (6.32) and (6.33) produce superior forecasts, in general, compared with other composite and individual forecasting techniques.

In this study, for the estimator (6.32), v was set equal to 1, 2, and 4. By using an MSE comparison, the composite estimates using (6.33) proved inferior to the USDA initial estimates for all three supply and inventory categories (see Tables 6.5, 6.6, and 6.7). Although inferior to the initial USDA estimates, the composite forecast of the breeding herd inventory with v = 2 provided a fairly accurate forecast. In Figure 6.10 weights on the initial USDA breeding herd inventory estimate are provided for this composite forecasting technique. The weights on the initial USDA breeding herd inventory estimate are given from the third quarter in 1970 to the fourth quarter in 1986. The USDA initial weight estimates seem to be the lowest when there is a temporary upturn in liquidation or an upturn during a breeding herd buildup. For example, during a breeding herd inventory buildup in 1977, the breeding herd temporarily moved downward by 4.5 percent between the June and December reports. The weight on the December report's initial estimate was only 0.11, and the RE and FME weights were 0.52 and 0.37, respectively. For the entire sample, however, the movements in the USDA weight are not due entirely to this sort of forecast error.

The forecast from the estimator (6.33) did appreciably better. By using an MSE comparison, the composite breeding herd inventory estimates were more accurate than the initial USDA estimates for v = 2 and α = 0.5 and 0.7. The USDA weight estimates for v = 2 and α = 0.5 are provided in Figure 6.11. The USDA weights seem to follow a cyclical pattern. In general, the USDA weights are greatest during peaks and troughs, and they drop during the gradual liquidation and buildup phases. Thus, similar to the results in Figure 6.10, the USDA weights decline when temporary upturns (downturns) occur during the liquidation (buildup) of the breeding herd. The forecast statistics for the complete set of alternative time-varying weight

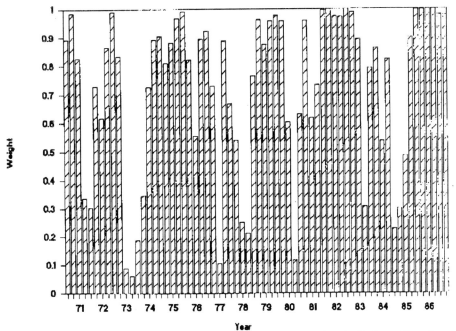

Figure 6.10. USDA initial estimate weight in composite U.S. breeding herd inventory forecast from estimator (6.32), v = 2

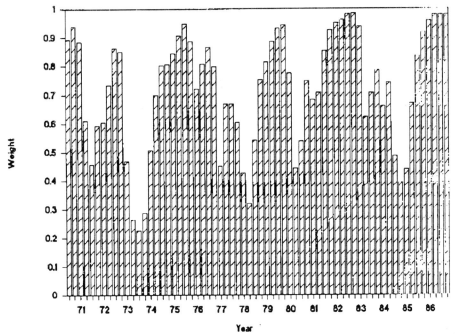

Figure 6.11. USDA initial estimate weight in composite U.S. breeding herd inventory forecast from estimator (6.33), v = 2, α = 0.5

Table 6.3. OLS estimation results of the USDA final estimates on the
nonstochastic fixed-weight composite forecasts

Dependent Variable	Intercept	Slope	Explanatory Variable	R^{2a}	D.W.[b]	Equation
BHUS	39.48 (0.41)[d]	0.99 (86.5)	$\hat{C}(6.14)$[c]	0.99	1.06	(6.34)
	72.78 (0.76)	0.99 (86.5)	$\hat{C}(6.29)$	0.99	1.05	(6.35)
FARROW	20.11 (0.38)	0.99 (57.8)	$\hat{C}(6.15)$	0.98	1.55	(6.36)
	59.30 (1.13)	0.98 (57.8)	$\hat{C}(6.30)$	0.98	1.54	(6.37)
PCUS	99.88 (0.32)	0.99 (73.4)	$\hat{C}(6.16)$	0.99	1.11	(6.38)
	271.96 (0.89)	0.99 (73.4)	$\hat{C}(6.31)$	0.99	1.11	(6.39)

[a] R^2 is the multiple correlation coefficient.
[b] D.W. is the Durbin-Watson d statistic.
[c] $\hat{C}(\cdot)$ is the composite forecast using the weights in the ($\cdot$) equation.
[d] In the parentheses is the t-ratio for the estimated coefficient.

Table 6.4. OLS estimation results of composite forecasts

Dependent Variable (Restriction)	Intercept	RE	FME	USDA	R^{2a}	D.W.[b]	Equation
BHUS (None)	52.89 (0.48)	-0.02 (-0.28)[c]	0.19 (1.77)	0.90 (21.3)	0.99	1.05	(6.40)
BHUS ($\beta_1+\beta_2+\beta_3=1$)	-5.43 (-0.45)	-0.0055 (-0.08)	0.11 (1.74)	0.89 (21.6)	0.99	1.06	(6.41)
BHUS ($\beta_0=0$)		-0.0068 (-0.10)	0.11 (1.74)	0.89 (21.6)	N/A	N/A	(6.42)
FARROW (None)	28.00 (0.45)	0.095 (0.77)	0.012 (0.11)	0.88 (17.9)	0.98	1.54	(6.43)
FARROW ($\beta_1+\beta_{2+}\beta_3=1$)	-5.30 (-0.69)	0.12 (1.12)	-0.0026 (-0.02)	0.88 (18.1)	0.98	1.54	(6.44)
FARROW ($\beta_0=0$)		0.12 (0.11)	-0.0063 (0.11)	0.88 (18.1)	N/A	N/A	(6.45)
PCUS (None)	127.2 (0.36)	0.052 (0.53)	0.043 (0.46)	0.90 (25.9)	0.99	1.11	(6.46)
PCUS ($\beta_1+\beta_2+\beta_3=1$)	-33.9 (-0.76)	0.071 (0.78)	0.031 (0.35)	0.99 (26.1)	0.99	1.09	(6.47)
PCUS ($\beta_0=0$)		0.067 (0.74)	0.033 (0.37)	0.90 (26.2)	N/A	N/A	(6.48)

[a] R^2 is the multiple correlation coefficient.
[b] D.W. is the Durbin-Watson d statistic.
[c] In the parentheses is the t-ratio of the estimated coefficient.

Table 6.5. Ridge regression estimation results of the composite forecasts

Dependent Variable	RE	FME	USDA	R^{2a}	D.W.[b]	Equation
BHUS	(-0.014)	0.12	0.89	N/A	N/A	(6.51)
	(-0.20)	(1.83)	(21.5)			
FARROW	0.053	0.043	0.90	N/A	N/A	(6.52)
	(0.54)	(0.47)	(26.1)			
PCUS	0.095	0.020	0.87	N/A	N/A	(6.53)
	(0.84)	(0.19)	(18.2)			

[a]R^2 is the multiple correlation coefficient.
[b]D.W. is the Durbin-Watson d statistic.
[c]In the parentheses is the t-ratio of the estimated coefficient.

estimates for the three supply and inventory categories are provided in Tables 6.6, 6.7, and 6.8.

Stochastic Methods

The stochastic composite forecasting methods assume that other random factors can influence the composite predictions of the final USDA estimates. Simply put, a random disturbance is included in the estimation of the weights of the competing forecasts, and the USDA final estimate is a random variable. The stochastic methods are a simple and direct method of obtaining the appropriate composite forecast weights.

The stochastic methods applied to the composite predictions of the breeding herd inventory, sows farrowing, and pig crop include applying ordinary least squares (OLS) to the composite estimates from the nonstochastic fixed-weight methods. Also, OLS is applied, with and without linear parameter constraints, to the USDA final estimates on the competing forecasts. Finally, ridge regression is used to determine the weights on the individual forecasts. Ridge regression has been suggested for composite

Table 6.6. Forecast statistics for the U.S. breeding herd inventory composite forecasts, 1970-86

| Method | Measure[a] | | | | |
	RMSE	RMPSE	MAE	MSE Ratio	Equation
USDA initial estimate	103.41	1.19	71.14		
Nonstochastic					
$\hat{k}_o(\epsilon_{BH})$	97.31	1.13	72.42	0.89	(6.14)
$\hat{k}_b(\hat{V}_{BH})$	97.39	1.13	71.13	0.89	(6.29)
$\hat{k}_i$ v=1	141.33	1.66	94.53	1.87	(6.32)
v=2	107.28	1.28	79.35	1.08	
v=4	107.76	1.27	81.49	1.09	
$\hat{k}_i$ v=1 α=0.5	124.51	1.46	90.42	1.45	(6.33)
α=0.7	110.06	1.33	86.45	1.13	
α=0.9	113.50	1.38	91.61	1.20	
v=2 α=-0.5	102.35	1.22	78.90	0.98	
α=0.7	103.13	1.23	81.30	0.99	
α=0.9	104.12	1.25	84.94	1.02	
v=4 α=0.5	106.91	1.27	83.43	1.07	
α=0.7	107.57	1.28	85.90	1.08	
α=0.9	104.23	1.25	84.58	1.02	
Stochastic					
$\hat{C}(\epsilon_{BH})$	97.01	1.12	71.89	0.88	(6.34)
$\hat{C}(\hat{V}_{BH})$	96.97	1.12	71.35	0.88	(6.35)
OLS	97.01	1.12	71.36	0.88	(6.40)
OLS restricted	97.02	1.13	71.54	0.88	(6.41)
OLS no constant	97.16	1.13	71.41	0.88	(6.42)
HKB-ridge	96.97	1.12	71.64	0.88	(6.51)

[a]RMSE is the root mean square error (6.1); RMPSE is the root mean percent square error (6.2); MAE is the mean absolute error (6.3); and MSE ratio is the MSE divided by the USDA initial estimate MSE.

Table 6.7. Forecast statistics for the U.S. sows farrowing inventory composite forecasts, 1970-86

Method	Measure[a]				
	RMSE	RMPSE	MAE	MSE Ratio	Equation
USDA initial estimate	64.3	2.27	36.16		
Nonstochastic					
$\hat{k}_o(\hat{\epsilon}_{FAR})$	61.24	2.16	35.59	0.91	(6.15)
$\hat{k}_b(\hat{V}_{BH})$	61.48	2.17	36.21	0.91	(6.30)
$\hat{k}_i$ $v=1$	83.62	2.62	45.59	1.69	(6.32)
$v=2$	74.84	2.56	45.15	1.35	
$v=4$	70.94	2.46	44.69	1.22	
$\hat{k}_i$ $v=1$ $\alpha=0.5$	82.14	2.55	47.75	1.63	(6.33)
$\alpha=0.7$	67.92	2.35	44.75	1.12	
$\alpha=0.9$	70.63	2.47	48.87	1.21	
$v=2$ $\alpha=-0.5$	70.04	2.44	45.07	0.19	
$\alpha=0.7$	68.56	2.43	45.18	0.14	
$\alpha=0.9$	66.48	2.36	42.88	1.07	
$v=4$ $\alpha=0.5$	69.57	2.42	45.99	1.17	
$\alpha=0.7$	69.29	2.44	45.77	1.16	
$\alpha=0.9$	67.88	2.39	44.41	1.11	
Stochastic					
$\hat{C}(\hat{\epsilon}_{FAR})$	60.91	2.15	36.30	0.91	(6.36)
$\hat{C}(\hat{V}_{B4})$	60.88	2.15	35.87	0.90	(6.37)
OLS	60.88	2.17	35.88	0.90	(6.43)
OLS restricted	61.02	2.17	36.61	0.90	(6.44)
OLS no constant	60.97	2.18	36.30	0.90	(6.45)
HKB-ridge	60.89	2.15	35.98	0.90	(6.52)

[a]RMSE is the root mean square error (6.1); RMPSE is the root mean percent square error (6.2); MAE is the mean absolute error (6.3); and MSE ratio is the MSE divided by the USDA initial estimate MSE.

Table 6.8. Forecast statistics for the U.S. pig crop composite forecasts, 1970-86

Method	Measure[a]				
	RMSE	RMPSE	MAE	MSE Ratio	Equation
USDA initial estimate	382.02	1.77	236.66		
Nonstochastic					
$\hat{k}_o(\hat{\epsilon}_{PC})$	358.51	1.66	244.59	0.88	(6.15)
$\hat{k}_b(\hat{V}_{PC})$	356.40	1.64	244.88	0.88	(6.30)
$\hat{k}_i$ $v=1$	464.69	2.07	282.95	1.48	(6.32)
$v=2$	467.61	2.03	283.55	1.50	
$v=4$	412.40	1.88	270.64	1.17	
$\hat{k}_i$ $v=1$ $\alpha=0.5$	397.64	1.75	278.40	1.08	(6.33)
$\alpha=0.7$	376.60	1.70	275.02	0.97	
$\alpha=0.9$	383.46	1.80	276.47	1.01	
$v=2$ $\alpha=-0.5$	401.03	1.80	265.23	1.10	
$\alpha=0.7$	387.47	1.73	262.88	1.03	
$\alpha=0.9$	382.80	1.79	263.82	1.00	
$v=4$ $\alpha=0.5$	413.89	1.89	282.22	1.17	
$\alpha=0.7$	413.01	1.92	289.08	1.17	
$\alpha=0.9$	410.02	1.91	286.55	1.15	
Stochastic					
$\hat{C}(\hat{\epsilon}_{PC})$	356.40	1.64	244.88	0.87	(6.38)
$\hat{C}(\hat{V}_{PC})$	356.31	1.64	223.55	0.87	(6.39)
OLS	356.30	1.64	223.57	0.87	(6.46)
OLS restricted	356.91	1.65	225.71	0.87	(6.47)
OLS no constant	356.67	1.65	225.41	0.87	(6.48)
HKB-ridge	356.30	1.64	223.66	0.87	(6.53)

[a]RMSE is the root mean square error (6.1); RMPSE is the root mean percent square error (6.2); MAE is the mean absolute error (6.3); and MSE ratio is the MSE divided by the USDA initial estimate MSE.

forecasting applications because of the near multicollinearity among the competing forecasts (Guerard 1987).

Ordinary Least Squares

The composite predictions from the nonstochastic, fixed-weight estimates, which assume the USDA final estimate is fixed, can be placed in a stochastic setting by applying OLS. The USDA final estimates are regressed on the nonstochastic, fixed-weight, composite estimates to obtain a new composite prediction. This new composite prediction allows other random factors, unexplained by the competing forecasts, to enter into the determination of the new estimate. The composite predictions derived from both forms of the estimated error variance-covariance matrix were used to form new composite predictions. Thus, the composite predictions for the three supply and inventory categories, derived from the weights in (6.14)-(6.15), and (6.29)-(6.31), were fitted with OLS against the USDA final estimate. The estimation results for these regressions are given in Table 6.3. The slope coefficients for all supply and inventory categories are statistically equal to one, and the statistically insignificant at conventional levels. The nonstochastic composite forecasts, based on R^2, explain most of the variation in the actual supply and inventory categories.

Granger and Ramanathan (1984) suggest obtaining the composite weight estimates by applying OLS directly, regressing the actual value on the individual forecasts with a constant term and no linear constraints on the parameters. They contend that this will produce an unbiased composite estimate, even if the individual forecasts are biased. The intercept term captures any biases in the individual forecasts. Of the composite techniques they tried, unrestricted OLS estimates produced the fewest in-sample MSE.

The unrestricted OLS estimation results for the composite forecasts of the breeding herd inventory, sows farrowing, and pig crop are presented in Table 6.4.

The values of the estimated weights are similar to the nonstochastic weight estimates. F tests of the hypothesis that composite weights (slope coefficients) sum to one cannot be rejected for any of the three models. Nearly all coefficients except the coefficient on the USDA initial estimate are insignificant at conventional levels. However, the FME estimate in the breeding herd inventory composite equation has a positive weight that is significant at the 5 percent level. The coefficient on the RE estimate of the breeding herd inventory has a negative sign. All composite forecasts explain a large proportion of the variation in the actual estimates.

Clemen (1986) suggests applying OLS directly, with linear parameter constraints to obtain the weight estimates. The linear constraint forces the slope coefficients to sum to one. Clemen contends that, if the individual forecasts are considered unbiased, the parameter constraints must be imposed. Also, if the individual forecasts are nearly unbiased, the parameter constraints also should be imposed to gain inefficiency, even though it may produce slight bias in the results. Also, depending on the nature of the individual forecasts, he suggests imposing the constraint that the intercept is zero.

The composite equations were estimated with the slope terms constrained to sum to one, with an intercept included and separately, without a constant. Given the results of the previous F tests, the constraints will not produce biased results and will improve the estimates' efficiency. As with other estimation results, in all of the unconstrained versions of the composite forecast, the USDA initial estimate receives by far the greatest weight. The coefficient signs and significance levels are similar to previous results. With the restricted equations, however, the sign on the FME estimate of sows farrowing becomes negative, but it still is not significantly different from zero.

Ridge Regression

The individual forecasts are highly correlated. This suggests the presence of near multicollinearity among the explanatory variables in the composite forecast regressions. Near multicollinearity gives rise to unstable parameter Table 6.4 estimates. It becomes more difficult to interpret the coefficients' contribution to explaining the dependent variable as the degree of collinearity increases. Also, the variance of the estimated coefficients increases sharply with the degree of collinearity. A common approach to the multicollinearity problem is ridge regression, which produces biased estimates of coefficients with smaller variances than the OLS estimator. The hope is that the induced bias is offset by the smaller variances, such that the ridge estimator MSE is reduced below that of OLS.

The ridge regression estimator is

$$\beta(\theta) = (X'X + \theta I)^{-1} X'Y , \qquad (6.49)$$

where

X = the N * K matrix of explanatory variables,

Y = N * 1 vector of the dependent variable,

I = K * K identity matrix,

β = K * 1 vector of the ridge parameter estimates, and

θ = a constant that is greater than zero (Judge et al. 1980).

The constant θ is usually called the shrinkage parameter. In most applications it lies between zero and one. When $\theta = \theta$, the OLS estimator is obtained. It can be shown that for some $\theta > \theta$, the MSE of $\beta(\theta)$ is less than the MSE of the OLS estimator (Judge et al. 1980, 474). However, $\beta(\theta)$ improves upon the OLS estimator only for certain ranges of θ, and the range of improvement depends on the unknown parameter $\beta(\theta)$ and σ^2, the variance of the disturbance term.

A method often used to search for the optimal θ is plotting ridge traces that are plots of the ridge estimates $\beta(\theta)$ and the residual sum of squares for different values of θ. The shrinkage parameter θ is chosen by assessing the tradeoff between coefficients' stability, size, and sign and the magnitude of the residual sum of squares. A more direct approach used in this study is the Hoerl-Kennard-Baldwin (HKB) estimator (Hoerl, Kennard, and Baldwin 1975). The NKB estimator of θ is defined as

$$\hat{\theta} = \frac{(K - 1)\, s^2}{\hat{\beta}'_c\, \hat{\beta}_c}, \tag{6.50}$$

where $\hat{\beta}_c$ is the OLS estimate obtained by regressing the centered dependent variable against the centered and standardized explanatory variables and s^2 is the estimated residual variance (see Judge et al. 1980, 474-75).

However, using any data-based search approach such as the HKB estimator to find the appropriate θ results in estimates with unknown reliability. The shrinkage parameter θ is a function of Y, a stochastic variable, and thus is a stochastic variable. The multicollinearity problem is addressed, but at the cost of losing a reliable statistical inference about the estimated coefficients. Also, there is no guarantee that the ridge MSE will be less than the OLS MSE.

The results of the HKB ridge regression procedure for the composite forecasts are presented in Table 6.5. The weights on the individual forecasts are quite similar to previous results. In general, the weights on the RE and FME are slightly less than previous results. However, the significance of the coefficients cannot be assessed with much confidence.

Composite Forecast Accuracy

The USDA initial estimates received the most weight in the composite forecasts. The RE and FME estimates received considerably less weight and were often insignificant in the composite estimates of the USDA final supply and inventory categories. This suggests that informational content of the USDA initial estimates is high relative to the competing RE and FME models in predicting the final estimates. However, the relative magnitude of the weights on the competing forecasts does not preclude the possibility of improving in the initial estimates in the *Hogs and Pigs* report.

Breeding Herd Inventory

The forecast statistics for the breeding herd inventory are summarized in Table 6.6. For convenience, the forecast statistics for the USDA initial estimate also are provided at the top of the table. All of the fixed-weight methods applied outperform the USDA initial estimates, according to an MSE criterion. The forecast performance for the stochastic methods is nearly identical. The ratio of the forecast MSE to the USDA initial estimate MSE was 0.88 to 0.89 for all fixed-weight methods. Also, in all fixed-weight methods, the RE estimate receives a negative weight. The weights of the FME and USDA initial estimates have similar values for all the fixed-weight techniques. The FME weights are significantly different from zero in the OLS (6.40) and in the restricted OLS versions (6.41 and 6.42).

The lowest MSE was obtained for the ridge regression estimates (6.51). The MSE for the composite forecast estimate based on the regressing nonstochastic forecast that allowed for individual forecast bias (6.35) on the final USDA estimate was nearly identical to the ridge regression estimates. In general, differences among the fixed-weight composite forecast statistics are negligible.

The composite methods that allow the weights to evolve through time show some promise as another means to improve the *Hogs and Pigs* report estimates of the breeding herd inventory. In particular, by using the estimator of k (6.33) with $v = 2$ and $\alpha = 0.5$ or 0.7, a prediction slightly better than the USDA initial estimates was obtained. However, the time-varying weight techniques were inferior to the fixed-weight methods. The MSE ratio ranges from 0.98 to 1.87, and only two are less than one.

Sows Farrowing

The forecast statistics for the composite forecasts and the USDA initial estimate are presented in Table 6.7. The results are similar to the breeding herd inventory composite forecasts. All fixed-weight methods outperform the USDA initial estimates. The MSE ratios for the fixed-weight estimates are 0.90 to 0.91. With the sows farrowing composite forecasts, the time-varying weight methods are inferior to the fixed-weight composite forecasts and to the USDA initial estimate in all instances.

The simple OLS regression, with a constant and no parameter restrictions (6.43), and the ridges estimates (6.52) provided the least forecast MSE. The MSEs for the other fixed-weight method are nearly identical to these two approaches. The relative sizes of the coefficients are quite similar across techniques, but the FME estimate receives a negative weight in the restricted OLS estimates [(6.44) and (6.45)]. The RE weight is always greater than the FME weight, and the USDA initial estimate weight always has a value of about 0.90. The contributions of the other competing forecasts are always insignificant.

Pig Crop

The performance of the pig crop composite estimates is similar to the previous categories. The forecast statistics for the pig crop composite estimates are provided in Table 6.8, with the measures of forecast performance of the USDA initial estimate. All of the fixed-weight composite estimates dominate the USDA initial estimate. The ridge regression estimate (6.53) provides the lowest MSE. The MSEs for the other stochastic methods are slightly greater, but only by trivial amounts. The ratio of the forecast MSE to the USDA initial estimate MSE ranges from 0.87 to 0.88 for the stochastic methods.

Nearly all of the time-varying weight methods are inferior to the USDA initial estimate. The time-varying parameter estimator (6.33) with $v = 2$ and $\alpha = 0.9$ produce a forecast with similar, but slightly higher, forecast statistics than the USDA initial estimate. The MSE ratio for this estimate was 1.00. And for the estimator (6.33) with $v = 1$ and $\alpha = 0.7$, an improved forecast of the pig crop is obtained. The MSE ratio for this estimate is 0.97. The MSE ratios for the time-varying parameter estimates ranges from 0.97 to 1.50.

Summary

Most of the composite predictions of the breeding herd inventory, sows farrowing, and the pig crop outperform the USDA initial estimates. In fact, all of the fixed-weight estimation methods used outperform the USDA initial estimates for all three supply and inventory categories. This implies that, even though the USDA initial estimates explain most of the variability in the final estimates in the *Hogs and Pigs* report, predictions from the market models of the pork sector still contain information useful to USDA officials.

Both econometric models use market information, available at the time the initial estimates were set. The RE and FME models are based on historical, biological, and behavioral relationships in the pork sector. The RE and FME models incorporate known biological relationships in their supply components. The models also incorporate alternative forms of expectations. The RE model uses an expectation mechanism that is consistent with the structure of the modeled pork sector. The FME model uses the expectations of futures market participants as proxies for the profitability anticipations of pork producers.

The market information, synthesized with the alternative econometric models, could be incorporated into estimation procedures in developing the initial estimates in the *Hogs and Pigs* report. Clearly, USDA officials could improve the accuracy of the initial estimates by incorporating the market information from the alternative expectation models with simple composite forecasting techniques. The time-varying weight methods show some promise in capturing the behavior of the breeding herd inventory. The fixed-weight composite forecasting methods provide clearly superior forecasts to USDA initial estimates.

Chapter 7

Implications and Conclusions

The assertion made at the outset was that the precision and consistency of the USDA estimate of the hog supply and inventory categories in the *Hogs and Pigs* report could be improved by expanding the information set beyond the current survey of agricultural producers. The rationale is clear. Fiscal constraints have reduced the survey coverage and have renewed concerns about the reliability and adequacy of the survey-based estimates.

In previous chapters, alternative information systems were developed to augment the survey-based estimates with market information. Specifically, the alternative information systems are econometric models of the pork sector. These models incorporate known biological constraints that govern the growth processes of hogs. The constraints form a consistency between the short-run stock-to-flow and flow-to-flow movements in the supply categories and the long-run formation of supply. The econometric models also include alternative mechanisms to represent individuals' expectations. The rational expectation model assumes that the anticipations of individuals are consistent with the behavioral structure of the pork sector. In the alternative model, futures market prices of live hogs and corn are used as proxies for the expectations of pork producers in their breeding herd management decisions.

These alternative systems provide a means to represent the often complex behavioral and structural relationships that affect the supply of hogs. The market information from these market models is synthesized through the one-step-ahead predictions of key hog supply and inventory categories. These predictions of the U.S. breeding herd inventory, sows farrowing, and pig crop

151

are merged with the USDA initial estimates with composite forecasting techniques. As shown in Chapter 6, the fixed-weight composite predictions clearly outperform the initial USDA estimates when evaluated by the mean square error criterion.

Forecast Contributions of the Alternative Systems

The competing information systems are assessed by comparing the percentage changes between the generated predictions and the actual supply or inventory categories. Thus, the percentage change of the predicted values from the composite forecast, rational expectations model, futures market expectation model, and the USDA initial estimates are compared with the USDA final estimates for the U.S. breeding herd, sows farrowing, and the pig crop.

In the *Hogs and Pigs* report, the USDA initial estimates are reported biannually for the aggregate United States in the June and December reports. As noted previously, the USDA final estimates are only available through 1982. Therefore, only the aggregate U.S. predictions for the June and December estimates for 1970-82 are used to compare the actual and predicted percentage change. This avoids errors due to data interpolations and recognizes that the USDA initial estimates are not yet finalized for the 1983-86 portion of the sample. Thus, the data used to obtain the actual percentage change for the breeding herd inventory are the final estimates of the June 1 and December 1 hogs kept for breeding (USDA 1977a; 1980; 1984). Similarly, for the actual percentage change of the sows farrowing and pig crop estimates, the data are the final estimates of the December-to-May and June-to-November sows farrowing and pig crop, respectively (USDA 1977a; 1980; 1984). The quarterly estimates were aggregated because quarterly initial estimates were not reported for the pig crop until 1978. The data for USDA initial estimates were obtained from the *Hogs and Pigs* report (USDA 1970-88).

The composite, rational expectation (RE), and futures market (FME) predictions are one-step-ahead predictions of the given supply or inventory category. The quarterly predictions of sows farrowing and the pig crop were aggregated to match the definition of the USDA actual and initial estimates. The composite predictions used in the percentage change comparisons for the breeding herd inventory, sows farrowing, and pig crop were obtained from the composite prediction with the least mean square error. For the breeding herd composite prediction, this was the ridge regression composite forecast (6.51). The sows farrowing composite forecast used the OLS estimate (6.44), and the pig crop used the ridge regression composite estimate (6.53).

Breeding Herd Inventory

The percentage changes indicated by the actual and competing predictions for the U.S. breeding herd inventory are provided in Table 7.1 for 1970-82. Comparisons of percentage change for the various individual forecasts with the actual movement in the breeding herd indicate the merit of combining market information in the estimation process of the *Hogs and Pigs* report.

The market information often counterbalances the overprediction of the initial estimate, thus resulting in more accurate composite predictions. For example, in December 1970, the initial estimate indicated more than a 12 percent drop in the breeding herd inventory from the previous June report. The RE and FME predictions indicated that the crop would be less, so the composite estimate was much closer to the finalized number. Also, during an upward movement in the breeding herd in June 1976, the initial estimate underpredicted the upward movement, but the composite estimate was much

Table 7.1. Percentage change of U.S. breeding herd inventory for final, composite, RE, FME, and USDA initial estimates

Year	Month	Final	Composite	RE	FME	Initial
			(percentage change)			
1970	June	15.68	19.17	18.04	16.26	19.65
	Dec.	-9.27	-11.06	-7.16	-2.58	-12.08
1971	June	1.07	4.13	3.28	-1.45	4.89
	Dec.	-13.06	-13.01	-10.84	-7.05	-13.81
1972	June	7.93	7.82	0.88	0.96	8.66
	Dec.	-5.43	-2.85	-2.91	-2.83	-2.88
1973	June	3.91	1.52	0.46	4.91	1.07
	Dec.	-4.26	-4.26	-1.20	-1.83	-4.55
1974	June	2.53	2.43	2.18	0.32	2.72
	Dec.	-16.25	-16.44	-9.16	-13.91	-16.72
1975	June	-0.42	-0.83	-8.54	-7.10	-0.15
	Dec.	2.94	2.71	-1.43	-0.88	3.09
1976	June	10.75	8.65	13.90	15.85	7.95
	Dec.	-4.49	-2.78	-2.51	-5.33	-2.45
1977	June	8.45	7.11	5.93	8.53	6.95
	Dec.	-0.97	1.58	-0.42	-1.51	1.94
1978	June	2.94	0.74	0.84	3.61	0.39
	Dec.	8.45	8.16	3.96	4.34	8.63
1979	June	7.94	9.27	13.63	16.31	8.53
	Dec.	-6.97	-7.99	-4.11	-8.19	-7.92
1980	June	-1.70	-0.66	-6.38	-4.88	-0.24
	Dec.	-3.83	-4.02	-3.38	-5.67	-3.82
1981	June	-8.34	-7.76	-1.81	1.43	-8.83
	Dec.	-6.15	-6.27	-5.06	-7.45	-6.12
1982	June	-5.48	-5.88	-8.07	-7.13	-5.80
	Dec.	0.82	-1.24	-3.31	-3.01	-1.07

closer to the actual change in the breeding herd inventory. Similar effects of combining market information can be found in other reports.

Again, the market models seem to counter estimate errors in the initial estimate. Only in two instance, December 1979 and 1980, did including market information cause greater errors in the composite estimate than in the USDA initial estimate. The RE and FME are less accurate predictors of inventory changes, but still provide useful information. The FME model provides better predictions than the RE market mode. At times, the FME prediction dominates the USDA initial estimate. For example, in June 1970, 1977, and 1978, the FME was more accurate than the USDA initial estimate. The RE predictions are clearly inferior. The RE predictions tend to overpredict the movements in the breeding herd. In part, this may be caused by the simple price determination structure of the model. The retail, wholesale, and farm price relationships may not be adequately captured by the simple structure posited.

Sows Farrowing

Similar results, as expected, are found in the individual and composite predictions in comparison with the actual percentage change in sows farrowing. Sows farrowing is closely related to underlying change in the breeding herd inventory. The percentage change of the individual, composite, and USDA final estimates for sows farrowing is provided in Table 7.2 for the 1970-82 period.

Including market information often compensates for the over- and underpredictions of the initial estimates. The market information in the RE and FME predictions is more noisy, compared with the initial predictions, but it often captures the direction of error in the initial estimates. For example, in June 1974 the USDA initially predicted that sows farrowing would increase

Table 7.2. Percentage change of U.S. sows farrowing for final, composite, RE, FME, and USDA initial estimates

Year	Month	Final	Composite	RE	FME	Initial
			(percentage change)			
1970	June	23.71	23.27	13.59	13.07	25.09
	Dec.	-3.25	-2.93	3.43	3.40	-3.75
1971	June	5.25	4.59	3.85	3.48	4.72
	Dec.	-12.41	-12.32	-8.54	-8.86	-12.90
1972	June	2.51	4.03	-0.42	0.00	4.68
	Dec.	-8.08	-4.86	-6.89	-5.70	-4.63
1973	June	7.79	4.24	6.92	5.51	3.93
	Dec.	-8.84	-10.03	-8.12	-3.97	-10.39
1974	June	7.60	8.77	8.04	6.77	8.95
	Dec.	-13.20	-13.43	-7.47	-10.14	-14.33
1975	June	-9.19	-8.99	-4.41	-8.31	-9.71
	Dec.	-0.42	-1.25	-12.53	-10.07	0.49
1976	June	16.16	14.20	11.45	10.32	14.72
	Dec.	1.26	3.22	3.97	6.50	3.13
1977	June	3.42	3.55	5.49	2.59	3.34
	Dec.	-0.68	3.35	-0.11	2.81	3.83
1978	June	0.42	-3.10	7.12	6.26	-4.46
	Dec.	6.03	4.85	-2.79	-0.71	6.00
1979	June	12.16	12.50	12.40	14.26	12.60
	Dec.	2.03	2.30	6.45	5.81	1.76
1980	June	-1.26	0.07	-1.49	-3.13	0.30
	Dec.	-5.19	-6.58	-6.62	-6.67	-6.62
1981	June	-6.05	-5.19	-0.70	-0.02	-5.85
	Dec.	-2.67	-3.21	-5.90	-5.84	-2.84
1982	June	-9.64	-9.87	-3.22	-5.37	-10.87
	Dec.	3.88	2.67	-6.71	-5.20	4.16

nearly 9 percent. The USDA final estimate indicated only a 7.6 percent increase, a clear overprediction. In the next report, the initial USDA estimate overpredicted the drop in sows farrowing by more than 1 percent. The RE and FME predictions suggest that less variability in sows farrowing would exist between the release of those reports. Consequently, the composite prediction more accurately reflected the changes in sows farrowing. Similar results are found in other reports.

The relative contributions of the RE and FME predictions are mixed. Unlike the predictions of breeding herd inventory movements, the RE predictions do at times outperform the FME predictions and they are superior even to the composite predictions and initial estimates. The RE predictions provide the best indication of sows farrowing movements in December 1972 and 1973 and June 1979 and 1980. This is surprising because of the inferior performance of the RE prediction of the breeding herd inventory.

On certain occasions, including market information exacerbates the error. The composite prediction is inferior to the initial estimate, for example, in December 1978. The initial estimate accurately suggested a 6 percent increase in sows farrowing from the previous report. The market signals, synthesized by the RE and FME models, suggested a downward movement in sows farrowing. This resulted in a less precise composite prediction compared with the initial estimate. However, this was the exception rather than the rule.

Pig Crop

The percentage changes implied by the individual, composite, and USDA initial estimates, along with actual movements in the pig crop, are given in Table 7.3 for 1970-82. Again, the RE and FME predictions

Table 7.3. Percentage change of U.S. pig crop for final, composite, RE, FME, and USDA initial estimates

Year	Month	Final	Composite	RE	FME	Initial
			(percentage change)			
1970	June	23.65	23.18	10.49	11.05	24.94
	Dec.	-4.87	-4.51	4.23	3.19	-5.32
1971	June	4.70	4.23	4.02	3.64	4.29
	Dec.	-11.39	-11.69	-8.68	-9.00	-12.05
1972	June	3.30	5.16	-0.27	0.16	5.75
	Dec.	-9.41	-5.50	-7.03	-5.85	-5.44
1973	June	7.14	2.82	7.10	5.68	2.47
	Dec.	-8.95	-9.92	-8.27	-4.13	-10.35
1974	June	6.65	7.50	8.22	6.94	7.54
	Dec.	-13.04	-13.23	-7.62	-10.28	-13.80
1975	June	-8.79	-8.85	-4.29	-8.20	-9.24
	Dec.	0.35	-0.01	-12.65	-10.18	1.37
1976	June	18.29	14.90	9.09	7.96	15.67
	Dec.	0.10	2.61	4.25	6.81	2.36
1977	June	1.76	2.02	6.26	3.33	1.74
	Dec.	0.56	4.69	0.29	3.23	5.04
1978	June	-1.67	-5.16	7.73	6.88	-6.45
	Dec.	8.36	7.22	-2.35	-0.27	8.26
1979	June	9.82	10.60	12.96	14.86	10.32
	Dec.	3.34	3.33	6.93	6.27	2.99
1980	June	0.09	1.29	-1.21	-2.84	1.66
	Dec.	-5.46	-6.73	-6.06	-6.12	-6.83
1981	June	-3.70	-3.23	-0.51	0.23	-3.58
	Dec.	-2.85	-3.15	-5.31	-5.30	-2.92
1982	June	-10.10	-10.54	-3.18	-5.33	-11.30
	Dec.	4.90	3.98	-6.23	-4.50	5.13

counteract the initial estimate errors. Over- and underpredictions of initial estimates are counterbalanced by the predictions of the RE and FME models.

The accuracy of the individual and composite predictions is mixed. In the 1973 December report, the RE prediction was the most accurate. In the next report, the FME prediction dominated the other forecasts. In the next two reports, the composite prediction was the most accurate. Finally, in the June 1977 report, the initial estimate most accurately predicted the slight increase in sows farrowing.

Implications and Conclusions

The results enumerated here, combined with the composite forecasting results in Chapter 6, conclusively demonstrate that the assimilation of market information in the process of developing USDA initial estimates creates more accurate and consistent estimates of the key hog supply and inventory categories. The RE and FME predictions, although providing relatively noisy signals of supply and inventory movements, offset the initial estimate errors. This is a compelling reason for the USDA to incorporate market information in the *Hogs and Pigs* report estimates.

The value of the greater consistency and precision of the estimates is unclear. Given the diminished resources devoted to the collection and dissemination of agricultural data, however, the demand for the adoption of more cost-effective information systems exists. Survey samples are expensive relative to the cost of developing and maintaining econometric models. Thus, the cost of improvement through adopting econometric market models is minimal compared with that of expanding existing survey coverage.

However, in the decision to alter the current information system, the cost and benefits of doing so must be determined (Miller 1977). Adopting market models in the data evaluation process requires an adequate quantification of the benefits and costs to justify its worth. This is a natural direction for future research. Earlier results, however, have indicated the worth of extra expenditures on public outlook information. Hayami and Peterson (1972) found that, from 1966 to 1968, the social loss associated with a 2.5 percent sampling error in the hog market was $1 million and that the social value of each additional dollar spent in reducing the sampling error from 2.5 to 2.0 percent for the major crop and livestock commodities was $600. If their results provide even a rough indication of the value of additional investment, adopting market models for evaluation and estimation is of merit.

Appendix A. Data Description and Sources

Variable	Label	Units	Source
Endogenous variables			
Additions to the breeding herd	ABHUS	1,000 head	$BHUS_t - BHUS_{t-1} + SSUS_t$
Sow slaughter	SSUS	1,000 head	USDA (1970-1986c)
Breeding herd inventory	BHUS	1,000 head	Reported biannually. December 1 (first quarter) and June 1 (third quarter) are obtained from USDA (1977a, 1980, 1984, 1970-1988) second- and fourth-quarter values are interpolations from ten-state data that are reported quarterly. $BHUS_t = BHUS_{t-1} * (BH10_t/BH10_{t-1})$, where BH10 is the ten-state values.
Sows farrowing	FARROW	1,000 head	USDA (1977a; 1980; 1984; 1970-88)
Pig crop	PCUS	1,000 head	USDA (1977a; 1980; 1984; 1970-88)
Barrow and gilt slaughter	BGSUS	1,000 head	USDA (1970-1986c)
Liveweight of barrows and gilts	LWBG	pounds	USDA (1970-83) and personal correspondence
Liveweight of sows	LWS	pounds	USDA (1970-83) and personal correspondence
Domestic pork production	PPF	pounds	$BGSUS_t * LWBG_t + SSUS_t * LWS_t$

161

Variable	Label	Units	Source
Commercial pork production	TOTSPK	million pounds	USDA (1970-1986c)
Domestic desappearance	TOTDPK	million pounds	USDA (1970-1986c)
Retail price of pork	RPPK	dollars per pound	USDA (1970-1986c) Divided by the CPI
Farm price of barrows and gilts	FPPK	dollars per pound	Barrows and gilts - 7 markets (USDA 1970-1986c)
Retail-farm margin	MARGIN	dollars per pound	$RPPK_t - FPPK_t/CPI_t$
Per capita pork consumption	PCPK	pounds per person	$(TOTDPK_t/POP_t) *$ $PVERT_t$

Exogenous variables

Variable	Label	Units	Source
U.S. population	POP	millions	U.S. Department of Commerce (1970-1986b)
Demand minus supply	OTHER	millions	TOTDPK - TOTSPK
Retail-carcass conversion	PVERT	USDA (1970-1986c)	
Feed costs	FC	dollars per bushel	(6/7) * (corn price/0.56) + (1/7) * (soybmeal price/20) Corn price data are from USDA (1970-1986a) and soybean meal price data are from USDA (1970-1986b)
Retail beef price	RPBF	dollars per bushel	Retail price of beef divided by the CPI (USDA 1970-1986c)
Per capita food expenditure	FEXP	dollars per person	Unseasonally adjusted food expenditure (U.S. Department of Commerce, personal correspondence) divided by POP and CPI.

Variable	Label	Units	Source
Marketing costs	MKTCST	1967 = 100	One-half of the index of meat packers' hourly earnings (U.S. Department of Commerce 1970-1986a), plus one-half of the producer price index of fuel and related power (U.S. Department of Commerce 1970-1986b), divided by the CPI.
Futures market price of live hogs	FUTHOG	dollars per hundredweight	Quarterly average of closing live hog futures prices (*The Wall Street Journal* 1970-86).
Futures market price of corn	FUTCORN	dollars per bushel	Quarterly average of corn futures prices (*The Wall Street Journal* 1970-86).
Quarterly dummy variable	D1, D2, D3, D4		
Dummy Variable	DL74		If year < 1974 equals one; equals zero otherwise
	DUM76		If year ≥ 1976 equals one; equals zero otherwise
	DUM73		If year ≥ 1973 equals one; equals zero otherwise
	D794		If year ≥ third-quarter 1979 equals one; equals zero otherwise
Time trend	T65		T65 = 1.00, 1.25, ...
Logarithm of time trend	LT65		Log (T65)

Variable	Label	Units	Source
Forecasted variables			
Rational expectation one-step-ahead forecast of U.S. breeding herd inventory	REBH	1,000 head	
Rational expectation one-step-ahead forecast of U.S. sows farrowing	REFAR	1,000 head	
Rational expectation one-step-ahead forecast of the U.S. pig crop	REPC	1,000 head	
Futures market expectation one-step-ahead forecast of U.S. breeding herd inventory	FUTBH	1,000 head	
Futures market expectation one-step-ahead forecast of U.S. sows farrowing	FUTFAR	1,000 head	
Futures market expectation one-step-ahead forecast of the U.S. pig crop	FUTPC	1,000 head	
USDA initial estimate of U.S. breeding herd inventory	USDABH	1,000 head	Reported biannually in June and December *Hogs and Pigs* report (USDA 1970-88). March and September values for aggregate U.S. are interpolations from initial ten-state estimates (see BHUS)
USDA initial estimate of U.S. sows farrowing	USDAFAR	1,000 head	USDA (1970-1988)

Variable	Label	Units	Source
USDA initial estimate of U.S. pig crop	USDAPC	1,000 head	Reported biannually in June and December *Hogs and Pigs* report (USDA 1970-88). Before 1978 reported as pig crop born in previous six months. Thus, before 1978, quarterly birth distribution assumed to follow final ten-state pig crop estimates

$$USDAPC_t = Initial *$$

$$\left(\frac{PC10_t}{PC10_t + PC10_{t-1}} \right),$$

where Initial is the initial estimate of the U.S. pig crop born in the previous six months and PC10 is the final ten-state pig crop estimate. Beginning in 1978, pig cro births are reported by quarter for the entire United States

REFERENCES

AAEA Committee on Economic Statistics. 1972. Our obsolete data systems: New directions and opportunities. *American Journal of Agricultural Economics* 54:867-875.

Almon, S. 1965. The distributed lag between capital appropriations and expenditures. *Econometrica* 33:176-178.

Aradhyula, S. V., and S. R. Johnson. 1987. Discriminating rational expectation models with non-nested hypothesis testing: An application to the beef industry. Working Paper 98-WP 19. The Center for Agricultural and Rural Development, Iowa State University, Ames, Iowa.

Arrow, K.S. 1962. Economic welfare and the allocation of resources for invention. In: *The Rate and Direction of Inventive Activity: Economic and Social Factors*. Princeton, New Jersey: Princeton University Press.

Arzac, E. R., and M. Wilkinson. 1979. A quarterly econometric model of the United States livestock and feed grain markets and some of its policy implications. *American Journal of Agricultural Economics* 61:297-308.

Askari, H., and J. T. Cummings. 1977. Estimating agricultural supply response with the Nerlove model: A survey. *International Economic Review* 18:257-292.

Bates, J.M., and C. W. J. Granger. 1969. The combination of forecasts. *Operations Research Quarterly* 20:451-468.

Begg, D. K. H. 1982. *The Rational Expectations Revolution in Macroeconomics*. Baltimore, Maryland: Johns Hopkins University Press.

Bessler, D. A., and J. A. Brandt. 1979. *Composite Forecasting of Livestock Prices: An Analysis of Combining Alternative Forecasting Methods*. Station Bulletin No. 265. Department of Agricultural Economics, Agricultural Experiment Station, Purdue University, West Lafayette, Indiana.

Blanton, B. 1983. A quarterly econometric model of the United States pork subsector. Unpublished M.S. thesis. University Missouri-Columbia.

167

Blanton, B., S. R. Johnson, J. A. Brandt, and M. T. Holt. 1985.
Applications of quarterly livestock models in evaluating and revising inventory data.
In: *Applied Commodity Price Analysis, Forecasting, and Market Risk Management*.
Proceedings of the NCR-134 Conference, Chicago, Illinois.

Board of Governors of the Federal Reserve System. 1982. *Agricultural
Finance Databook*. Quarterly series E.15 (125). Division of Research and Statistics,
Washington, D.C.

Bonnen, J. T. 1977. Assessment of the current agricultural data base: An
information system approach. In: G. G. Judge, R. H. Day, S. R. Johnson, G. C.
Rausser, and L. R. Martin, eds. *A Survey of Agricultural Economics Literature*.
Vol. 2. Minneapolis, Minnesota: University of Minnesota Press.

_____. 1983. The dilemma of agricultural economists: Discussion.
American Journal of Agricultural Economics 62:889-890.

Bonnen, J. T., and G. L. Nelson. 1981. Changing rural development data
needs. *American Journal of Agricultural Economics* 63:337-345.

Bottum, J. C., and J. Ackerman. 1958. Current and area data progress and
future needs in the United States. *Journal of Farm Economics* 50:1772-1778.

Box, G. E. P., and G. M. Jenkins. 1976. *Time Series Analysis: Forecasting
and Control*,. San Francisco: Holden-Day.

Bradford, D. F., and H. H. Kelejian. 1977. The value of information for
crop forecasting in a market system: Some theoretical issues. *Review of Economic
Studies* 44:519-531.

_____. 1978. The value of information for crop forecasting with
bayesian speculators: Theory and empirical results. *The Bell Journal of Economics*
9:123-144.

Brandt, J. A., and D. A. Bessler. 1981. Composite forecasting: An
application with U.S. hog prices. *American Journal of Agricultural Economics*
63:135-140.

_____. 1983. Price forecasting and evaluation: An application in
agriculture. *Journal of Forecasting* 2:237-248.

_____. 1984. Forecasting with vector autoregressive versus a
univariate ARIMA process: An empirical example of the U.S. hog prices. *North
Central Journal of Agricultural Economics* 6:29-36.

Brandt, J. A., R. Perso, S. Alam, R. E. Young II, and A. Womack. 1985.
Documentation of the CNFAP Hog-Pork Model and Review of Previous Studies.
CNFAP Staff Report, CNFAP-9-85. Center for National Food and Agricultural
Policy, University of Missouri-Columbia.

Bray, M. 1981. Futures trading, rational expectations, and the efficient market hypothesis. *Econometrica* 49:575-596.

Breeden, D. T. 1980. Consumption risks in futures markets. *Journal of Finance* 35:503-520.

Brennan, M. J. 1959. The supply of storage. *American Economic Review* 48:50-72.

Bullock, J. B. 1976. Social costs caused by errors in agricultural production forecasts. *American Journal of Agricultural Economics* 58:76-80.

_____. 1981. Some concepts for measuring the value of rural data. *American Journal of Agricultural Economics* 63:346-352.

Cagan, P. 1956. The monetary dynamics of hyper-inflation. In: M. Friedman, ed., *Studies in the Quantity of Money*. Chicago: University of Chicago Press.

Chavas, J.-P., and S. R. Johnson. 1982. Rational expectations in econometric models. In: G. C. Rausser, ed., *New Directions in Econometric Modeling and Forecasting in U.S. Agriculture*. New York: Elsevier-North Holland.

Chavas, J.-P., R. D. Pope, and R. S. Kao. 1983. An analysis of the role of futures prices, cash prices and government programs in acreage response. *Western Journal of Agricultural Economics* 8:27-33.

Chavas, J.-P., and R. M. Klemme. 1986. Aggregate milk supply response and investment behavior on U.S. dairy farms. *American Journal of Agricultural Economics* 68:55-66.

Clemen, R. T. 1986. Linear constraints and the efficiency of combined forecasts. *Journal of Forecasting* 5:31-38.

Clemen, R. T., and R. L. Winkler. 1986. Combining economic forecasts. *Journal of Business and Economic Statistics* 4:39-46.

Coase, R. H., and R. F. Fowler. 1937. The pig-cycle in Great Britain: An explanation. *Economica* 4:55-82.

Cochrane, W. W. 1966. Improvements needed in statistics for making policy and program decisions. *Journal of Farm Economics* 48:1654-1666.

Cootner, P. H. 1960. Returns to speculators: Telser vs. Keynes. *Journal of Political Economy* 68:396-404.

Cromarty, W. A. 1959. An econometric model for United States Agriculture. *Journal of the American Statistical Association* 54:556-574.

Dean, G. W., and E. O. Heady. 1958. Changes in supply response and elasticity for hogs. *Journal of Farm Economics* 40:845-860.

Doan, T. A., and R. B. Litterman. 1987. *User's Manual RATS.* Version 2.10. Evanston, Illinois: VAR Econometrics, Inc.

Draper, N., and H. Smith. 1981. *Applied Regression Analysis.* New York: John Wiley and Sons.

Dusak, K. 1973. Futures trading and investor returns: An investigation of commodity market risk premiums. *Journal of Political Economy* 81:1387-1406.

Eckstein, Z. 1984. A rational expectations model of agricultural supply. *Journal of Political Economy* 92:1-19.

Eisgruber, L. M. 1973. Managerial information systems in the U.S.A.: Historical development, current status, and major issues. *American Journal of Agricultural Economics* 55:930-939.

Ezekiel, M. 1938. The cobweb theorem. *Quarterly Journal of Economics* 52:255-280.

Fair, R. C. 1980. Estimating the uncertainty of policy effects in nonlinear models. *Econometrica* 48:1381-1391.

_____. 1984. *Specification, Estimation, and Analysis of Macroeconometric Models.* Cambridge, Massachusetts: Harvard University Press.

Fair, R. C., and J. B. Taylor. 1983. Solution and maximum likelihood estimation of dynamic nonlinear rational expectations models. *Econometrica* 51:1169-1185.

Falk, B., and P. Orazam. 1986. A theory of market response to government crop forecasts. Staff paper series No. 150. Department of Economics, Iowa State University, Ames, Iowa.

Fama, E. F. 1970. Efficient capital markets: A review of theory and empirical work. *Journal of Finance* 25:383-417.

Fama, E. F., and K. R. French. 1987. Commodity futures prices: Some evidence on forecast power, premiums, and the theory of storage. *Journal of Business* 60:55-73.

Federal Reserve Bank of Chicago. 1983-1986. *Agricultural Letter.* Chicago, Illinois.

Feldstein, M. S. 1971. The error of forecast in econometric models when the forecast period exogenous variables are stochastic. *Econometrica* 39:55-59.

Ferris, J. 1962. Unsolved problems in data collection and analysis. *Journal of Farm Economics* 44:1763-1772.

Fomby, T. B., R. C. Hill, and S. R. Johnson. 1984. *Advanced Econometric Methods.* New York: Springer-Verlag.

Foote, R. J. 1953. A four-equation model of the feed-livestock economy and its endogenous mechanism. *Journal of Farm Economics* 35:44-61.

Fox, K. A. 1953. *The Analysis of Demand for Farm Products.* Technical Bulletin 1081. U.S. Department of Agriculture.

Freebairn, J. W., and G. C. Rausser. 1975. Effects of changes in the level of U.S. beef imports. *American Journal of Agricultural Economics* 57:676-688.

Futures. 1984. Much ado over hog reports. *Futures* (August):62.

Gardner, B. L. 1983. Fact and fiction in the public data budget crunch. *American Journal of Agricultural Economics* 65:882-888.

_____. 1976. Futures prices in supply analysis. *American Journal of Agricultural Economics* 58:81-84.

_____. 1975. The farm-retail price spread in a competitive food industry. *American Journal of Agricultural Economics* 57:399-409.

Goodwin, R. M. 1947. Dynamic coupling with special reference to markets having production lags. *Econometrica* 25:181-204.

Goodwin, T. H., and S. M. Sheffrin. 1982. Testing the rational expectations hypothesis in an agricultural market. *The Review of Economics and Statistics* 64:658-667.

Granger, C. W. J., and P. Newbold. 1986. *Forecasting Economic Time Series.* New York: Academic Press.

Granger, C. W. J., and R. Ramanthan. 1984. Improved methods of combining forecasts. *Journal of Forecasting* 3:197-204.

Grossman, S. J., and J. E. Stiglitz. 1980. On the impossibility of informationally efficient markets. *American Economic Review* 70:393-408.

Grundmeier, E., K. D. Skold, and S. R. Johnson. 1988. *CARD Livestock Model Documentation: Beef.* Technical Report 88-TR 2. The Center for Agricultural and Rural Development, Iowa State University, Ames, Iowa.

Guerard, J. B. 1987. Composite forecasting using ridge regression. *Communications in Statistics* 16:937-952.

Harlow, A. A. 1962. A recursive model of the hog industry. *Agricultural Economics Research* 14:1-12.

_____. 1960. The hog cycle and the cobweb theorem. *Journal of Farm Economics* 42:842-853.

Hayami, Y., and W. Peterson. 1972. Social returns to public information services: Statistical reporting of U.S. farm commodities. *American Economic Review* 62:119-130.

Hayenga, M., V. J. Rhodes, J. A. Brandt, and R. E. Deiter. 1985. *The U.S. Pork Sector: Changing Structure and Organization.* Ames, Iowa: Iowa State University Press.

Heady, E. O., and D. R. Kaldor. 1954. Expectations and error in forecasting agriculture prices. *Journal of Political Economy* 62:34-37.

Heien, D. 1975. An econometric model of the U.S. pork economy. *Review of Economics and Statistics* 57:370-375.

_____. 1977. Price determination processes for agricultural sector models. *American Journal of Agricultural Economics* 59:125-136.

Helmberger, P. G., G. R. Campbell, and W. D. Dobson. 1981. Organization and performance of agricultural markets. In: L. R. Martin, ed., *A Survey of Agricultural Economics Literature.* Vol. 3. Minneapolis, Minnesota: University of Minnesota Press.

Hicks, J. R. 1939. *Value and Capital.* Oxford: Clarendon.

Hieronymus, T. A. 1971. *Economics of Future Trading: For Commercial and Personal Profit.* New York: Commodity Research Bureau.

Hirshleifer, J. 1973. Where are we in the theory of information? *American Economic Review* 63:31-39.

Hoerl, A., R. Kennard, and K. Baldwin. 1975. Ridge regression: Some simulations. *Communications in Statistics* 4:105-123.

Hoffman, G. 1980. The effect of quarterly livestock reports on cattle and hog prices. *North Central Journal of Agricultural Economics* 2:145-150.

Hogs and Pig Report. n.d. Des Moines: U.S. Department of Agriculture.

Hohmann, K. 1987. Take a closer look at *Hogs and Pigs* reports. *Hog Farm Management* (March):40-43.

Holt, M. T., and S. R. Johnson. 1988. Bounded price variation, rational expectations, and endogenous switching in the U.S. corn market. Working Paper 88-WP 28. The Center for Agricultural and Rural Development, Iowa State University, Ames, Iowa.

Hudson, M., S. Koontz, and W. Purcell. 1985. Why hog futures react so wildly to USDA reports. *Futures* (March):60-61.

Huntzinger, R. L. 1979. Market analysis with rational expectations. *Journal of Econometrics* 10:127-145.

Ives, J. R. 1957. An evaluation of available data for estimating market supplies and prices of cattle. *Journal of Farm Economics* 39:1411-1418.

Jarvis, L. S. 1974. Cattle as capital goods and ranchers as portfolio managers: An application to the Argentine cattle sector. *Journal of Political Economy* 82:489-520.

Jessen, R. J. 1978. *Statistical Survey Techniques.* New York: John Wiley & Sons.

Johnson, S. R., R. D. Green, Z. A. Hassan, and A. N. Safyurtlu. 1986. Market demand functions. In: O. Capp, Jr., and B. Senauer, ed., *Food Demand Analysis: Implication for Future Consumption.* Department of Agricultural Economics, Virginia Polytechnic Institute and State University, Blacksburg, Virginia.

Johnson, S. R., and T. G. MacAulay. 1982. Physical accounting in quarterly livestock models: An application for U.S. beef. Working Paper. University of Missouri-Columbia.

Johnson, S. R., and G. C. Rausser. 1982. Composite forecasting in commodity systems. In: G. C. Rausser, ed., *New Directions in Econometric Modeling and Forecasting in U.S. Agriculture.* New York: North-Holland.

Jorgenson, D. W. 1966. Rational distributed lag function. *Econometrica* 34:135-149.

Judge, G. G., W. E. Griffiths, R. C. Hill, and T. Clee. 1980. *The Theory and Practice of Econometrics.* New York: John Wiley and Sons.

Just, R. E. 1983. The impact of less data on the agricultural economy and society. *American Journal of Agricultural Economics* 65:872-881.

Just, R. E., and G. C. Rausser. 1981. Commodity price forecasting with large-scale econometric models and the futures market. *American Journal of Agricultural Economics* 63:197-208.

Kaldor, N. 1939. Speculation and economic stability. *Review of Economic Studies* 7:1-27.

Keynes, J. M. 1930. *A Treatise on Money.* Vol. II of *The Applied Theory of Money.* London: Macmillan.

King, R. P. 1983. Technical and institutional innovation in North American grain production: The new information technology. In: C. F. Runges, ed., *The Future of the North American Grainary: Politics, Economics, and Resource Constraints in North American Agriculture.* Ames, Iowa: Iowa State University Press.

Koffsky, N. M. 1962. What the federal-state farm economic intelligence service is and does. *Journal of Farm Economics* 44:1754-1759.

Krog, D. R. 1988. Plant-process corn yield forecasts for Iowa. Unpublished Ph.D. dissertation, Iowa State University, Ames, Iowa.

Kutish, F. A. 1955. Needed changes in state and local crop and livestock reports. *Journal of Farm Economics* 37:1050-1053.

Leuthold, R. M. 1974. The price performance on the futures market of a nonstorable commodity: Live beef cattle. *American Journal of Agricultural Economics* 56:271-279.

Leuthold, R. M., and P. A. Hartman. 1981. An evaluation of the forward-pricing efficiency of livestock futures markets. *North Central Journal of Agricultural Economics* 3:71-80.

_____. 1979. A semi-strong form evaluation of the efficiency of the hog futures market. *American Journal of Agricultural Economics* 61:482-489.

Leuthold, R. M., P. Garcia, B. Adam, and W. I. Park. 1987. A re-examination of the pricing efficiency of the hog-futures market. In: *NCR Conference on Applied Commodity Price Analysis, Forecasting, and Market Risk Management.* Proceedings of the NCR-134 conference, Chicago, Illinois.

Luby, P. J. 1957. Evaluation of available data for estimating marketing supplies and prices of hogs. *Journal of Farm Economics* 39:1402-1410.

Lucas, R. E. 1976. Econometric policy evaluation: A critique. *Journal of Monetary Economics* 1(supplement):19-46.

MacAulay, T. G. 1978. A forecasting model for Canadian and U.S. pork sectors. In: *Commodity Forecasting Models for Canadian Agriculture*, Publication No. 7812, Policy and Economics Branch. Ottawa: Agriculture Canada.

McCallum, B. T. 1976. Rational expectations and the estimation of econometric models: An alternative procedure. *International Economic Review* 17:484-490.

Maki, W. R. 1962. Decomposition of the beef and pork cycles. *Journal of Farm Economics* 44:731-743.

Maki, W. R., C. Y. Liu, and W. C. Motes. 1962. *Interregional Competition and Prospective Shifts in the Location of Livestock Slaughter.* Iowa Agricultural Experiment Station Research Bulletin No. 511. Iowa State University, Ames, Iowa.

Marsh, J. M. 1977. *Effects of Marketing Costs on Livestock and Meat Prices for Beef and Pork.* Montana Agricultural Experiment Station Bulletin No. 697. Department of Agricultural Economics and Economics, Montana State University, Bozeman, Montana.

Meilke, K. D., A. C. Zwart, and L. J. Martin. 1974. North American hog supply: A comparison of geometric and polynomial distributed lag models. *Canadian Journal of Agricultural Economics* 22:15-30.

Metzler, L. A. 1941. The nature and stability of inventory cycles. *The Review of Economic Statistics* 23:113-29.

Meyer, S. R., and J. D. Lawrence. 1988. Comparing USDA *Hogs and Pigs* reports to subsequent slaughter: Does systematic error exist? In : *NCR conference on Applied Commodity Price Analysis, Forecasting, and Market Risk Management.* Proceedings of the NCR-134 conference, St. Louis, Missouri.

Miller, S. 1979. The response of futures prices to new market information: The case of live hogs. *Southern Journal of Agricultural Economics* 11:67-70.

Miller, S. E., and E. E. Kenyon. 1980. Empirical analysis of live-hog futures prices use by producers and packers. In: R. M. Leuthold and P. Dixon, ed., *Livestock Futures Research Symposium.* Chicago: Mercantile Exchange.

Miller, T. A. 1977. *Value of Information: A Project Prospectus.* Agricultural Policy Analysis Program Area, Commodity Economics Division, Economic Research Service, U.S. Department of Agriculture.

Moore, K. C. 1985. Predictive econometric modeling of the United States farmland market: An empirical test of the rational expectations hypothesis. Unpublished Ph.D. dissertation. Iowa State University, Ames, Iowa.

Moschini, G., and K. D. Meilke. 1988. Structural change in U.S. meat demand: further evidence. In: *NCR conference on Applied Commodity Price Analysis, Forecasting, and Market Risk Management.* Proceedings of the NCR-134 conference, St. Louis, Missouri.

Muth, J. F. 1961. Rational expectations and the theory of price movements. *Econometrica* 29:315-335.

Nelson, C. R. 1975. Rational expectations and the predictive efficiency of econometric models. *Journal of Business* 48:331-343.

Nerlove, M. 1956. Estimates of the elasticities of supply of selected agricultural commodities. *Journal of Farm Economics* 38:496-509.

_____. 1958. *The Dynamics of Supply: Estimation of Farmers' Response to Price*. Baltimore, Maryland: Johns Hopkins Press.

_____. 1979. The dynamics of supply: Retrospect and prospect. *American Journal of Agricultural Economics* 61:874-888.

Newbold, P., and C. W. J. Granger. 1974. Experience with forecasting univariate time series and the combination of forecasts. *Journal of the Royal Statistical Society* 137:131-146.

Okyere, W. A. 1982. A quarterly econometric model of the United States beef sector. Unpublished Ph.D. dissertation. University of Missouri-Columbia.

Okyere, W. A., and S. R. Johnson. 1987. Variability in forecasts in a nonlinear model of the U.S. beef sector. *Applied Economics* 19:1457-1470.

Oleson, F. H. 1987. A rational expectation model of the United States pork industry. Unpublished Ph.D. dissertation. University of Missouri-Columbia.

Pankratz, A. 1983. *Forecasting with Univariate Box-Jenkins Models*. New York: John Wiley and Sons.

Peck. A. 1976. Futures markets, supply response, and price stability. *Quarterly Journal of Economics* 90:407-423.

Phillip, D. O. A. 1986. Rational price expectations and structural change in the U.S. broiler market. Unpublished Ph.D. dissertation. Iowa State University, Ames, Iowa.

Pindyck, R. S., and D. L. Rubinfeld. 1981. *Econometric Models and Economic Forecasts*. New York: McGraw-Hill.

Quandt, R. E., and S. M. Goldfeld. 1987. *GQOPT4/I*, Version 4.02. Department of Economics, Princeton University, Princeton, New Jersey.

Rausser, G. C., and C. Carter. 1983. Futures market efficiency in the soybean complex. *Review of Economics and Statistics* 65:469-478.

Rosen, S. 1987. Dynamic animal economics. *American Journal of Agricultural Economics* 69:545-557.

Shideed, K. H., and F. C. White. 1988. Time-varying weighting schemes for the combination of forecasts: An application to supply response of U.S. soybean acreage. In: *Applied Commodity Price Analysis, Forecasting, and Market Risk Management*. Proceedings of the NCR-134 conference, St. Louis, Missouri.

Shonkwiler, J. S. 1982. An empirical comparison of agricultural supply response mechanisms. *Applied Economics* 14:182-194.

Shonkwiler, J. S., and R. D. Emerson. 1982. Imports and the supply of winter tomatoes: An application of rational expectations. *American Journal of Agricultural Economics* 64:634-641.

Simpson, G. D. 1966. Resources and facilities for providing needed statistics: The role of the statistical reporting service. *Journal of Farm Economics* 48:1674-1682.

Skold, K. D., E. Grundmeier, and S. R. Johnson. 1988. *CARD Livestock Model Documentation: Pork*. Technical Report 88-TR 4. The Center for Agricultural and Rural Development, Iowa State University, Ames, Iowa.

Skold, K. D., and M. T. Holt. 1988. Dynamic elasticities and flexibilities in a quarterly model of the U.S. pork sector. In: *NCR Conference on Applied Commodity Price Analysis, Forecasting, and Market Risk Management*. Proceedings of the NCR-134 conference, St. Louis, Missouri.

Stein, J. L. 1981. Speculative price: Economic welfare and the idiot of chance. *Review of Economics and Statistics* 63:223-232.

Stillman, R. 1985. *A Quarterly Model of the Livestock Industry*. Technical Bulletin No. 1711. Economic Research Service, U. S. Department of Agriculture.

Stopher, P. R., and A. H. Meyburg. 1979. *Survey Sampling and Multivariate Analysis for Social Scientists and Engineers*. Lexington, Massachusetts: D. C. Heath and Company.

Subonik, A., and J. P. Houck. 1982. A quarterly econometric model for corn: A simultaneous approach to cash and futures markets. In: G. C. Rausser, ed., *New Directions in Econometric Modeling and Forecasting in U.S. Agriculture*. New York: North-Holland.

Tegene, A., W. E. Huffman, and J. A. Miranowski. 1988. Dynamic corn supply functions: A model with explicit optimization. *American Journal of Agricultural Economics* 70:103-111.

Telser, L. G. 1958. Futures trading and the storage of cotton and wheat. *Journal of Political Economy* 66:233-255.

_____. 1967. The supply of speculative services in wheat, corn, and soybeans. *Food Research Institute Studies* 8(supplement):131-76.

Tomek, W. G., and R. W. Gray. 1970. Temporal relationships among prices on commodity future markets. *American Journal of Agricultural Economics* 52:372-380.

Tomek, W. G., and K. L. Robinson. 1977. Agricultural price analysis and outlook. In: L. R. Martin, ed., *A Survey of Agricultural Economics Literature*, Vol. 1. Minneapolis: University of Minnesota Press.

_____. 1981. *Agricultural Product Prices.* 2nd ed. Ithaca, New
York: Cornell University Press.

Trelogan, H. C., C. E. Caudill, H. F. Huddleston, W. E. Kibler, and E.
Brooks. 1977. Technical developments in agricultural estimate methodology. In:
G. G. Judge, R. H. Day, S. R. Johnson, G. C. Rausser, and L. R. Martin, ed., *A
Survey of Agricultural Economics Literature*, Vol. 2. Minneapolis, Minnesota:
University of Minnesota Press.

Tullock, G. 1970. *Private Wants, Public Means.* New York: Basic Books.

U.S. Bureau of the Census. 1967. *United States Census of Agriculture 1964.*
Washington, D.C.: U.S. Government Printing Office.

_____. 1978. *United States Census of Agriculture 1974.*
Washington, D.C.: U.S. Government Printing Office.

_____. 1981. *United States Census of Agriculture 1978.*
Washington, D.C.: U.S. Government Printing Office.

_____. 1984. *United States Census of Agriculture 1982.*
Washington, D.C.: U.S. Government Printing Office.

U.S. Department of Agriculture. 1970-1986a. *Agricultural Prices.*
Washington D.C.: National Agricultural Statistics Service, Agricultural Statistics
Board.

_____. 1986. *Agricultural Statistics.* Washington, D.C.: U.S.
Government Printing Office.

_____. 1970-1986b. *Feed Situation and Outlook.* Washington,
D.C.: Economic Research Service.

_____. 1977a. *Hogs and Pigs Final Estimates for 1970-75.*
Statistical Bulletin No. 588. Washington D.C.: Statistical Reporting Board, Crop
Reporting Board.

_____. 1980. *Hogs and Pigs Final Estimates for 1976-78.*
Economics and Statistics Service, Crop Reporting Board, Statistical Bulletin No. 648.
Washington, D.C.: U.S. Government Printing Office.

_____. 1984. *Hogs and Pigs Final Estimates for 1979-82.*
Statistical Reporting Service, Crop Reporting Board, Statistical Bulletin No. 716.
Washington, D.C.: U.S. Government Printing Office.

_____. 1970-88. *Hogs and Pigs.* Washington, D.C.:
National Agricultural Statistics Service, Agricultural Statistics Board.

_____. 1977b. Hog reports and market prices.
Agricultural Situation. Washington, D.C.: Statistical Reporting Service.

_____. 1970-1986c. *Livestock and Poultry Situation and Outlook.* Washington, D.C.: Economic Research Service.

_____. 1970-1983. *Livestock and Meat Statistics.* Washington, D. C.: Economic Research Service.

_____. 1961. *Pig Crops.* Agricultural Marketing Service, Crop Reporting Board, Statistical Bulletin No. 276. Washington, D.C.: Government Printing Office.

_____. 1983. *Scope and Methods of the Statistical Reporting Service.* Statistical Report Service, Miscellaneous Publication No. 1308. Washington, D.C.: U.S. Government Printing Office.

_____. 1947. *Sows Farrowing by Months.* Washington, D.C.: Bureau of Agricultural Economics, Crop Reporting Board.

_____. 1988. *Summary of 1988 Data Users Meeting.* Washington, D.C.: U.S. Department of Agriculture.

_____. 1982. *USDA's Statistical Reporting Service Changes: Crop and Livestock Estimating Program.* Washington, D.C.: Statistical Reporting Service.

U.S. Department of Commerce. 1970-1986a. *Employment and Earnings.* Washington, D.C.: Bureau of Labor Statistics.

_____. 1970-1986b. *Survey of Current Business.* Washington, D.C.: Bureau of Economic Analysis.

Upchurch, M. L. 1977. Developments in agricultural economic data. In: G. G. Judge, R. H. Day, S. R. Johnson, G. C. Rausser, and L. R. Martin, ed., *A Survey of Agricultural Economics Literature*, Vol. 2. Minneapolis, Minnesota: University of Minnesota Press.

Van Arsdall, R. N., and K. E. Nelson. 1984. *U.S. Hog Industry.* Agricultural Economic Report No. 511. Washington, D.C.: U.S. Department of Agriculture, Economic Research Service, U.S. Government Printing Office.

Wallis, K. F. 1980. Econometric implications of the rational expectation hypothesis. *Econometrica* 48:49-73.

The Wall Street Journal. 1970-1986.

Wickens, M. R. 1982. The efficient estimation of econometric models with rational expectations. *Review of Economic Studies* 59:55-67.

Wilkinson, M. 1985. Futures prices as embedded forecasts: The case of corn and livestock. Working Paper Series #CSFM-107. Center for the Study of Futures Markets, Columbia University.

Winkler, R. L., and S. Makridakis. 1983. The combination of forecasts. *Journal of the Royal Statistical Association* 146:150-157.

Wohlgenant, M. K., and J. D. Mullen. 1987. Modeling the farm-retail price spread for beef. *Western Journal of Agricultural Economics* 12:119-25.

Working, H. 1942. Quotations on commodity futures as price forecasts. *Econometrica* 10:39-52.

_____. 1948. The theory of the inverse carrying charge in futures markets. *Journal of Farm Economics* 30:1-28.

Zanias, G. P. 1987. Adjustment costs and rational expectations: An application to a tobacco export model. *American Journal of Agricultural Economics* 69:22-29.

INDEX

A

Accuracy, forecast, 7-8, 113, 116-20, 134, 146-49
Agricultural Finance Databook, 90
Agricultural Letter, 90
Agricultural Prices, 90
ARIMA, 84, 95-99, 103, 114, 115
Assessment, forecast, 152-59

B

Barrows, 62, 71, 72
Bias, 7, 108, 114, 143, 144
  boundary definition, 23
  nonresponse, 31
  variance-covariance matrix and, 129-32
Biology
  in econometric models, 58-59, 63-66, 90-94
  of livestock production, 5, 6-7
  of pork production, 7, 58-63, 90-94
Boars, 73
Boundary definitions, 23
Breeding herd, 62
  additions to, 69, 78
  biological impact on, 91-92
  data sources, 89-90
  disinvestment in, 69
  estimates, 6, 7, 37
    composite, 153-55
    final, 116. *See also* Estimates, USDA final
    initial, 114, 115. *See also* Estimates, USDA initial
    revised, 38
    standard error for, 32-33
    weights on, 134, 140, 143
  forecasts, 114, 115, 116, 120, 134, 138
    assessment of, 152-55
    errors, 117-19
    statistics, 147
  improving inventory of, 147
  slaughter relationship to, 6-7, 63
  variance-covariance in forecast of, 127-28, 130-32

C

Cattle on Feed, 4-6
CBT. *See* Chicago Board of Trade
Census of Agriculture, 15, 19, 33, 38, 116
Chicago Board of Trade (CBT), 77, 90
Chicago Mercantile Exchange (CME), 77, 90
CME. *See* Chicago Mercantile Exchange
Cobweb theorem, 57
Commodity reports, 4
Computers, 1-2, 3, 104
Corn Belt, 17
Cost
  econometric models versus surveys, 159-60
  feed, 69, 77, 115
  sampling, 22, 25

D

Data
  collecting, 2
  cutbacks in coverage of, 4
  evaluation with alternative systems, 11-12
  slaughter, 33
  sources, 89-90
  system criticism, 4
Database, 2-4
Davidson-Fletcher-Powell (DFP) method, 87-89
Demand, 59, 73-76
  futures market expectation model, 105
  rational expectation model, 104
DFP. *See* Davidson-Fletcher-Powell method
Doanes Agricultural Service, 2

E

Elasticities, supply, 110-11
Employment and Earnings, 90
EP. *See* Extended Path method
Error
  forecast, 11, 44-45, 115, 116, 117-24, 126
  Hogs and Pigs report, 8
  reducing, 11
  relative standard, 32-33
  sampling and nonsampling, 8, 32
  sources of, 8